GOLD!

MADNESS, MURDER, AND MAYHEM IN THE COLORADO ROCKIES

IAN NELIGH

WESTWINDS PRESS®

Library of Congress Cataloging-in-Publication Data

Names: Neligh, Ian Paul, author.
Title: Gold! : madness, murder, and mayhem in the Colorado Rockies / Ian Paul Neligh.
Description: Portland, Oregon : WestWinds Press, an imprint of Graphic Arts Books, 2017. | Includes bibliographical references. |
Identifiers: LCCN 2017012535 (print) | LCCN 2017033907 (ebook) | ISBN 9781513260655 (paperback) | ISBN 9781513260679 (hardbound) | ISBN 9781513260662 (ebook)
Subjects: LCSH: Colorado—Gold discoveries—History. | Gold miners—Rocky Mountains—Biography. | Gold mines and mining—Rocky Mountains—History.
Classification: LCC F776.6 (ebook) | LCC F776.6 .N45 2017 (print) | DDC 978.8/02—dc23
LC record available at https://lccn.loc.gov/2017012535

Designed by Vicki Knapton

Cover Image Credits: skull: iStock.com/ianmcdonnell; background: wet2017/Shutterstock.com

Published by WestWinds Press®
An imprint of

GraphicArtsBooks.com

Contents

For Billie, of course.

Introduction

Gold! Clear as day and twice as bright, the glittering piece of metal winked up at me from the tide pool of black sand. Hardly daring to breathe, I adjusted my pan again and coaxed another little wave to further reveal the treasure. The dirt drew back, and my heart began beating faster. Mouth dry, I could hardly believe it. I'd found an actual piece of Colorado gold.

It was the same gold that tempted the Spanish to venture into dangerous new lands hundreds of years ago. It was the same gold that inspired legend, provoked madmen, dreamers, and treasure hunters. It was the same gold, even when found in the smallest quantities, that set fire to a gold rush that swept across the United States in 1859 and drew to the Rockies a staggering 100,000 people. Towns formed, laws were cast, and a state was born—all because of the same gold that I now gazed down at.

As a newspaper reporter, I'd spent years working in and around the towns and cities established in the desperate scramble for gold. I'd worked in the brick buildings, walked the narrow streets, and seen the amber-colored stains running from forgotten mines like the aftermath of bullet holes from a gunfight.

I'd looked at a bygone era's hulking relics, left to lean dangerously from the hillsides and valleys, rusted tributes to a time when a fortune could be dug from the ground and anyone, regardless of their economic status, could change it all in the blink of an eye. They also serve as memorials to crushed dreams, lives, and an environmental legacy that will chain us to the sites for all of time.

What I found more compelling were those that still hunted for their fortune in the shade cast by the gold rush more than 150 years ago. Prospectors, miners, and treasure hunters who ignored popular sentiment that the gold was gone, that it had disappeared or was too hard to remove. A small community engages in dangerous, backbreaking work even today to pry wealth from the dirt and rock of the Colorado mountains.

Fascinated with both the history of the gold rush and those who still toiled in its legacy, I spent a year meeting with them, hearing their stories, and trying to understand why it is they continue to do what they do—often in the face of extreme hardship. Many times by word of mouth, I went and met with one after the other and discovered the dubious inheritance of the gold rush included far more than just miners and prospectors.

But for the moment those thoughts were far away. I'd found gold in a chilly Colorado stream, and that is, after all, how it all began.

CHAPTER 1

WOLVERINES AND SUNKEN TREASURE

Despite the bitter cold, George Jackson continued wading through waist-deep snow, going ever farther west into what would become the Colorado Rockies. Originally from Missouri, the hunter, trapper, and experienced prospector had no clear destination; Jackson just wanted to see what was beyond the next bend in the river. In retrospect this was maybe not the best idea as he had nearly drowned some weeks before and was saved by one of his traveling companions. But now Jackson was alone, save for his two dogs, and often risked injury or death. Even so, he continued west. It was the winter of 1859.

On January 2, Jackson woke to hear his two dogs growling in the frigid blue hue of early morning. Eyes open, he scanned his campsite. The nearby herd of bighorn sheep he had spotted the day before were now gone. Kit and Drum continued their low, intense warning, which created plumes in the biting air. Then he spotted it. The mountain lion was only twenty feet away. The difference between life and death on the frontier was sometimes as simple as attacking first.

"[I] pulled my gun from under the blankets. Shot too quick; broke his shoulder," Jackson wrote in his diary. He fired again, the second gunshot report deafening in the mountain canyon. The lion dropped dead to the snow.

"Clear high wind and very cold," Jackson later remarked of the day, adding he spent this time in camp building with tree branches a small shelter from the freezing temperatures. The next day he spotted another mountain lion creeping up on him, which he also shot dead.

On January 4, Jackson and his dogs followed the river, which would later be named Clear Creek, for five miles, then followed the north fork of the river for five more miles through the rugged, ankle-splintering country. This was a land that had been seen briefly by the Spanish some two hundred years before, but was known to Native American tribes such as the Utes and Cheyenne. Exhausted, Jackson returned to his camp after dark and discovered yet another surprise.

"Mountain lion stole all of my meat in camp; no supper tonight—damn him."

Jackson didn't know it, but he would soon make a discovery at the confluence of two creeks that would send many thousands of settlers into this far-flung western portion of what was then the Kansas Territory. The call to fame and fortune would dwarf the size of the California gold rush, bringing in miners, merchants, entrepreneurs, criminals—and lead to the formation of a state, which today has some 5.5 million residents. In just one more day George Jackson would make a discovery so large, it would light the fuse that set off the Colorado gold rush.

Bottom of the Ocean

In a near-abandoned high school parking lot, just south of the historic city of Idaho Springs, sits a monument dedicated to George Jackson. A giant and unimaginative potato-shaped boulder rests on a pedestal, hidden to one side by a grove of small trees. A plaque fixed to its front reads:

> *"On this spot was made the first discovery of gold in the Rocky Mountains by George A. Jackson January 7th, 1859 placed 1909."*

Jackson's discovery wasn't the first in Colorado or even the largest—but it was the first time a substantial amount of gold was found in the Rockies. Before the high school in Idaho Springs was built, and later abandoned for a larger one; before neighboring Interstate 70 snaked its way up the canyon along Clear Creek, connecting the plains to the mountains; before even the town, the mills, and the mines that preceded

George Jackson.
(Courtesy of the Historical Society of Idaho Springs)

them all, Jackson, with his two dogs, fought their way deeper into a largely unexplored canyon.

There is some disagreement about Jackson's original intentions in the Rockies. While he was certainly a seasoned veteran of the California gold rush ten years before, his trip into the Rockies lacked any prospecting supplies and seemed to indicate he had come to Colorado mainly for hunting and trapping. A small amount of gold was discovered in Colorado only the year before, and rumors and legends of the precious metal had persisted since at least 1765 when Spanish explorer Don Juan María Antonio de Rivera returned from Colorado.

The Spaniard had brought samples of gold with him to Santa Fe, which were later dismissed by his government. Subsequent travelers, explorers, mountain men, and even madmen related tales of gold that were likewise disregarded. The California gold rush of 1849 saw those who were seeking to strike it rich cross through the Rockies and pan the streams along the way.

In 1850 Lewis Ralston, on his way to California, stopped for a time in Colorado to pan a small amount of gold from a drifting finger of Clear Creek. The gold was quickly removed from the area and he decided to move on, continuing his journey west. Again gold was discovered but in such small quantities that it didn't warrant additional time or energy.

"For some ten years past, vague stories affirming or implying the existence of gold in our country's principal chain of mountains, have from time to time reached the public ear; but they seemed to rest on very slight or insecure foundations, and attracted but limited and transient attention," wrote the *New York Tribune*'s Horace Greeley in his account from 1859. "An Indian's, or trapper's, or trader's bare assertion that, in traversing the narrow ravines and precipitous heights of our American Switzerland he had picked up a piece of quartz lustrous with gold, or even a small nugget of the pure metal, was calculated to attract little attention, while California was unfolding her marvelous treasures. . . ."

William Green Russell had participated in the gold rush in his home state of Georgia in the 1830s, and in California's a decade later. With his eye on the Rockies, Russell organized a party to prospect Colorado with the guidance of Cherokee Indians related to him through marriage. The Cherokee told him they had discovered gold in the streams that tumbled down from those breathtaking mountains. In July of 1858, Russell, his two brothers, and their small party finally succeeded in finding gold in Dry Creek. While they were able to pan out only a limited quantity of gold, it had been found, and word of its discovery began working its way back east.

In time, its battle cry would become the plucky and courageous, however geographically erroneous, "Pikes Peak or Bust."

Indeed, the Russells had looked for gold near Pikes Peak, far to the south of where they had discovered gold, and in a dozen other places trying to find the source and were as yet unsuccessful. However, the cry had gone out, and settlements and later towns began growing in the places they had searched. Soon the towns of Auraria and Denver came into being. But the gold the Russells had found was depleted and no one had yet struck it rich. Regardless, the stories of great wealth continued. A haphazard sprint for the newly discovered goldfields ensued.

"Though they carried home or sent home large stories of the

auriferous character of the country they 'prospected,' [and] took with them precious little gold," Greeley reported. "But their reports aroused a spirit of gold-seeking adventure in others, so that the ensuing (last) fall witnessed a rush of three or four hundred, mainly men of broken fortunes from the dead mushroom 'cities' of Nebraska and Kansas, to the region watered by the South Platte and the more northerly sources of the Arkansas [River]."

As it turns out, the country was also in a severe depression, starting with the Panic of 1857. Homes, businesses, jobs were all lost in the economic crises, a nasty situation primarily aggravated by the loss of the gold-laden SS *Central America* in a hurricane. The disappearance of the valuable cargo dealt a near-crippling blow to the American economy.

The ship sank near the Carolinas, taking with it more than four hundred passengers and twenty-one tons of gold to the bottom of the sea. Fortunately, many women and children were evacuated before the ship went down, and another fifty were later rescued from the ocean. A handful of men survived a desperate week in a lifeboat before being discovered.

Incidentally, the gold was later recovered in 1988 by treasure hunter Tommy G. Thompson, who was arrested in 2015 after a two-year manhunt for failing to appear before a judge in a case where investors were excluded from the gold profits removed from the wreck.

In the wake of the tragic incident with the ship, and compounding economic troubles, merchants across the Midwest who were desperate for any source of additional income were all too eager to help perpetuate the talk of gold and what could have easily become a myth.

"I doubt that three thousand dollars' worth of gold in every shape had been taken out by the five or six hundred seekers who came to this region in hot pursuit of it," Greeley wrote.

However, the gold was there: they only needed to look a little higher.

A Golden Ring

In the late days of 1858, Jackson was hunting with his friends Tom Golden and Black Hawk. His diary has brief descriptions of the terrible weather and his varied successes aiming down the length of his rifle.

> Dec. 27
> "Still snowing. Tom hunting Oxen. Black Hawk and I for elk. I killed a fine fat doe. Still snowing."
>
> Dec. 28
> "Snowing fast, accompanied by high wind. In camp all day."
>
> Dec. 29
> "I got into camp late at night; saw about 600 elks; killed five cows and one bull."
>
> Dec. 31
> "Jerked Elk meat until noon with intention of going down mountain . . . packed meat and blankets and started down over fallen timber and through snow four feet deep. Had a hell of a time before I reached the creek. Went into camp at dark. Dogs and I almost tired out. Made big fire after supper and dried my clothes and blankets. Turned in about 12 o'clock, and slept good until daylight."

Then on the first day of the new year, Jackson decided to head out on his own to follow the stream into the mountains. He told Tom Golden that he'd be back at their camp above Table Mountain in a week. With two pounds of bread, one pound of coffee, and dried elk for both himself and his dogs, Jackson set off.

Traveling about eight miles farther upstream, he killed a mountain lion he came across, and as some accounts of Jackson's adventures report, he saw what he initially interpreted as a cloud of smoke from a camp of Native Americans. Being cautious, he worked his way through the deep snow and saw that it was actually steam from a hot spring. The snow around it was melted away, and the sheep he came across were eating the thawed vegetation.

"Killed fat sheep and camped under three cottonwood trees. About 1,000 mountain sheep in sight tonight; no scarcity of meat in future for myself and dogs. Good," Jackson reported. He was up before daylight, shot at and wounded another mountain lion, and drank the last of his coffee. He then started inspecting the gravel of the streams. "Good gravel here; looks like it would carry gold," Jackson wrote. "Wind has blown snow off the rim, but gravel is hard frozen. Panned out two cups; nothing but fine colors."

The next day he built a giant fire on the rocks to thaw the gravel. He kept the fire going all day and didn't notice at first that a "carcajou," also known as a wolverine, had come into his camp. What followed was a savage fight between the dogs, Jackson, and the wolverine.

"Dogs killed him after I had broken his back with belt ax. Hell of a fight."

On January 7 Jackson removed the embers from the fire he had set the day before to thaw the frozen gravel by the stream. Using his cup, he panned out the gold from the rock and was quite pleased with what he found. On his ninth cupful, he found a large nugget of gold. He would later have it turned into a ring for his wife.

"Feel good tonight," he wrote in his journal, then added as an afterthought, "Carcajou no good for dog." He worked at the stream all the next day with the inadequate tools that he'd brought along. He made the best of what he had, even wearing out his belt knife.

"Well, Tom old boy, I've got the diggins at last," Jackson wrote. "But can't be back in a week. Dogs can't travel. Damn carcajou."

After recovering about an ounce of gold, he decided to head back to camp and join up with his hunting partners—that is once his dogs were ready to travel again. In preparation for a return trip, he carefully hid evidence of his work and marked a tree with a knife and his belt ax.

"Snowing like hell. High wind and cold. In camp all day. Drum can hardly walk around today." It did finally stop snowing, but Jackson spent the day in his camp doctoring his dog's leg, which he said had swollen to the size of his upper arm.

"Damn a carcajou."

On January 12 the three of them began the slow trip back down the mountains. That evening Jackson put a balsam on Drum's wound. They started late the next day but made better progress; he noted that "Drum is doing much better."

On January 14 with his moccasins so worn that he was nearly barefoot, Jackson made his way back to camp and found his friend Tom more than a little uneasy about his delay in returning.

"After supper, I told him what I had found and showed him the gold, and we talked, smoked and ate, the balance of the night. I could hardly realize I had been gone nineteen days." Once out of the mountains he and Tom came across a man they knew who was using two sluice boxes to get gold from the stream. Compared with what Jackson had pulled from the river in his coffee cup just days prior, he was not remotely impressed.

"No good; too fine to save without quicksilver, and not enough to pay with it."

Many miners used quicksilver, or mercury, to remove the gold from rock or to help remove it from sand. The mercury turns the gold into an amalgam, which can later be burned off and returned to its pure form. Not environmentally friendly or safe, but it is effective and is still used by some today. Jackson decided he wanted to return to his discovery not only with better supplies but with a small party of men to help him work the confluence of the two creeks. However, he would have to bide his time until spring.

"Tom is the only man who knows I found gold up the creek, and as his mouth is as tight as a No. 4 beaver trap, I am not uneasy."

The same month a party of prospectors discovered gold in nearby Boulder. Called Gold Hill, the men also knew it was essential to keep the news of their discovery quiet and hidden from prying eyes as long as possible. On April 17 of that same year, Jackson returned to the area he had marked with twenty-two men, wagons, supplies, and tools. The group often had to build their own road, hacking through the dense wilderness, which made for a grueling journey into the mountains. In some places, the wagons were unable to get through and so had to be meticulously disassembled and reassembled on the other side of each obstruction. This was slow and painful work. In May they reached Jackson's location and made $1,900 in the first seven days. It's said about $2.5 million in gold was removed from the area near his discovery in three years' time. Originally called Jackson's Diggings, the area provided the first real evidence that gold could be found and fortunes dug from the muddy gravel of the Rockies. This was not rumor or legend but undeniable truth.

In time Jackson felt himself called away from the mountains and their promise of wealth by the drums of war. Jackson fought in the Civil

War in 1861 for the Confederates. He did return to Colorado when the fighting was done to look for gold in Ouray, but his further attempts to strike it rich were ended suddenly one day when his firearm fell from a wagon and accidentally discharged, killing him.

Jackson's dreams of finding untold riches lived on. People from across the nation and even the oceans rushed into the area. Rough cabins and tents popped up like mushrooms over every free space in sight in the area that would soon become Idaho Springs. Some have estimated that 100,000 people joined the hunt for gold in 1859. Some 13,158 claims were recorded in Clear Creek County from the start of the gold rush to 1861. Supplies, food, and plenty of coffee and whiskey were paid for in gold dust. The streams were soon depleted of gold, and miners began to sink shafts alongside the creek banks to the bedrock below to find where gravity had carried the gold over time.

The hunt for gold naturally evolved to see where it came from and how it got in the streams. Soon miners were chasing the gold lodes or veins into the surrounding mountainsides, and with the bang of black powder, the era of hard rock mining began. Men drilled by hand blasting rock apart and used candlelight to work by.

Necessity is the mother of invention and pneumatic drills were created and black powder gave way to dynamite. By 1902 the Idaho Springs area had more than three hundred mines, which were estimated to have a combined one hundred miles of tunnels.

CHAPTER 2

A FORTUNE LOST AND FOUND

He tried looking again and still couldn't find it. Everything was covered in thick snow, and the source of what could have been a fortune in gold was now gone. It was April 1859, several months after George Jackson made the historic find at the confluence of Chicago Creek and Clear Creek. Jackson would soon head up again to remove the gold with his team—and John Gregory found himself in something of an unfortunate predicament.

Often characterized as ornery, cantankerous, and something of an antisocial curmudgeon, Gregory was a prospector in the truest sense. He was described as bearded, red shaggy-haired, wiry, wrinkled, and dressed as a beggar. One can also be fairly certain he didn't smell of flowers but much else about him is a mystery. Unlike Jackson, who happened accidentally upon his find, Gregory was specifically hunting the streams for gold. But gold dust or a simple handful of nuggets wasn't at all what he had in mind. Gregory had his eye on a larger prize. He wanted to find where the gold had come from.

John Gregory had left his home of Gordon County, Georgia, in August the year before with the intention of looking for gold in British Columbia. He was detained in his travels and forced to spend the winter in Fort Laramie. The breathless news of the discovery of gold in Colorado

John Gregory. *(Courtesy of the Gilpin Historical Society)*

reached him, and a classic opportunist, he decided immediately to change his plans. Gregory followed the gold flakes like bread crumbs in the mountain rivers and up the north fork of Clear Creek. A seasoned prospector, he knew exactly what he was looking for and soon found it.

A vein or lode of gold ore often streaks through the area where the precious metal has filled in a fissure in the rock. In time a small amount of gold washes away where it is exposed to the surface and runs into the streams. But Gregory knew that if he could indeed discover the source of the gold found in the streams, he'd find a fortune.

In what is essentially an overflow casino parking lot today, Gregory came across a ledge that had the state's most famous lode of gold ore. A sudden spring blizzard forced him to use his gold pan to dig a shelter and build around him a brush hut. He was stuck there for days, with his supply of food rapidly dwindling. When the storm finished, and he climbed out into the blinding winter landscape, he saw his discovery was gone. In the deep snow, he couldn't for the life of him find it again. One can imagine his frustration echoing in colorful verse off the surrounding valley walls.

With no supplies, no money, and no options, Gregory was forced to leave the state's first gold vein discovery and head back down the mountain to the town that would later be called Golden, after George Jackson's good friend.

Penniless, Gregory got to the town and it's said he took a little time to recover from his adventure through the restorative powers of the local saloon. Before long he found people who believed his discovery was real and would help him outfit a small party to journey into the mountains to rediscover his gold. On May 6, with a fresh group of treasure hunters from Indiana, Gregory marched forth once again to locate the spot that had eluded him. With luck, and a seasoned eye, he found the lost vein once again.

By then word of the rich gold discoveries of Colorado had already caught fire and were roaring east, back across the Great Plains. Journalist Henry Villard recalled that an exodus took place when word reached those living in the young towns of Denver and Auraria.

"Whoever could secure provisions enough for a stay in the mountains started off without delay," Villard wrote in his memoir. "Traders locked their stores, barkeepers set out with their stock of whiskey, the few mechanics [carpenters] that were engaged in building houses dropped their work. The county judge and sheriff, lawyers and doctors, and even the editor of the *Rocky Mountain News*, joined in the rush. Naturally, I did not stay behind, but started out on a fine mule. . . ."

When Villard reached the area where Gregory had found gold, he was exhausted from the long trip. He asked to be pointed in the direction of the prospector, found him, introduced himself, and "begged a place to lie down for the night. He complied at once, and assigned me a corner of his tent," Villard wrote. "My animal required no care, as he had had plenty of grass and water on the way, and, after picketing him, I spread my blankets and was asleep in a moment."

When Villard woke in the morning, he took a moment to get his bearings of the area he called the "Gregory Mine."

"Although but two weeks had elapsed since Gregory had washed out the first 'pay dirt' in his pan, there were already many scores of men busily engaged in ripping open the mountain sides with pick and shovel," Villard wrote. "Dozens of huts of pine branches had been erected, and tents pitched. Sluices, 'long toms,' and 'rockers' were in full operation, ditches crossed the gulch, and slides were being constructed—in short, the very picture of a busy, promising mining camp was before me."

Villard met with Gregory and interviewed him about his background. Gregory describes his group's early days being hindered by ice and snow.

"But for a week the weather has been warm enough. A great many, as you see, have tracked us to the gulch and taken up claims on other veins and are working them," Villard reported Gregory as saying. Villard then spent about a week with Gregory and the other miners looking for gold and enjoying their "hospitality."

"I visited every 'lead' and 'claim' then opened, witnessed the digging, hauling, and washing of 'pay-dirt,' washed out many a pan myself, saw the gold in the riffles of the sluices, and was daily present when the workers caught the quicksilver used to gather the fine gold from the sluices and heated it in retorts into gold-charged cakes," Villard wrote. "Thoroughly convinced by all this ocular evidence that the new [El] Dorado had really been discovered, I returned to Denver, and felt justified in spreading this great news with all the faith and emphasis of conviction."

Again ready to make history, William Green Russell showed up in the area on June 1, 1859, with more than one hundred followers ready to find gold. He found Gregory's area already filled to the brim with people also hoping to strike it rich. He continued on an additional two miles from the mine, also called "Gregory Gulch," and struck gold in what would later become "Russell Gulch."

The miners began sanctioning the area off into claims and mining districts. Then the towns began appearing: Central City, Black Hawk, Mountain City, Nevadaville, Gregory Point. Today only the first two still exist; history turned the others into ghost towns.

Gregory himself earned $972 in six days on his claim. But like Jackson, he wasn't destined to stay for long in the area he had discovered. Gregory sold his claim for $20,000 and in time disappeared, largely without a trace, from the pages of history.

Hung from a Yellow Pine

"[Gregory] was an experienced miner from Georgia, had been in the gold rush there and like a lot of people kind of heard what was going on in the West and thought he understood it," said David Forsyth, executive director and curator of the Gilpin Historical Society. "They had the placer gold discoveries . . . which kind of got people to looking. And so what he did is he followed it up Clear Creek."

I met with Forsyth in his office on the top floor of a giant 146-year-old schoolhouse converted into a museum. All around him and all the way to the ceiling were stacks of books and other yellowed reference materials. As we talked about Gregory, the whistling of the ancient boiler occasionally interrupted us.

"And within about six weeks of his discovery there were thirty thousand people up here," Forsyth said. "Because placer gold is nice—but you got to work a lot of placers to get enough gold to make it worth your while. That discovery was huge and with the people coming up here, staking their claims . . . it was chaotic at first."

Chaotic indeed. Russell reported in the first few months from his area that several men had already been shot, five froze to death, more drowned trying to ford a river, and eighteen died in various forest fires. A Capt. Wm. M. Slaughter recalled an incident where he and two friends were prospecting twenty miles northwest of Gregory when they came across a small party of Utes. Apparently, the men shared dinner and after the meal, the groups went their different ways—his friends to prospecting the streams and the Utes to hunting. When Slaughter later returned to the group, he was shocked at seeing the Indians busy scalping his friends. He hid among the rocks and made his way back to Gregory's diggings to share his dramatic tale. Crime and claims jumping had to be curtailed early on because they soon discovered that more money and investors were needed as the gold was chased into the hard rock.

"So very quickly they realized this wasn't going to work, so each mining district started making their own rules—and it was 'This is how we're going to handle claims, this is how we're going to handle claim jumpers, and this is how we're going to handle crime,'" Forsyth said. "If it was a certain crime, they might shave half your head and send you out of

A man pans for gold near where it was first discovered on Clear Creek.
(Photo by L. McLean, courtesy of the Historical Society of Idaho Springs)

town. They might tar and feather you. Claim jumping you could be killed for. They were not shy about it."

It was not even a full year later when the area had its first lynching when a man named Pensyl Tuck attempted to shoot Mountain City Sheriff Jack H. Kehler. The lawman was apparently quicker on that draw and shot and wounded Tuck.

"[Tuck had] gone to a miner's court, and he had threatened everyone involved with it," Forsyth said. "The trial adjourned, and Tuck tried to shoot the sheriff, who returned fire and hit Tuck. He was taken to his cabin, the doctor dressed his wounds." In what probably wasn't the cleverest move, Tuck told his physician that he planned on doing some killing in the name of revenge as soon as he was up and out of bed. Understandably concerned, the doctor decided to warn those men.

"Over the next few days he repeatedly threatened to kill basically everyone in Mountain City," Forsyth said. "Two hundred men approached his cabin, dragged him from his bed, and they hanged him from the limb of a nearby yellow pine. That was the first lynching in Gilpin County.

People objected to it, not because he didn't deserve to be hanged, but because they thought he should have had a trial first."

The territory's first "legal" execution also occurred in Central City in 1863 after William Van Horn killed a man out of jealousy when the girl he was with dumped him.

"They were very serious about these rules and regulations," Forsyth said. "Because they wanted these outside investors to come in and they realized that the easy lode gold was gone, and they were going to have to start doing hard rock mining. And you can't do that without money. Investors don't want to go to a place where there are shootouts in the street three times a day. They want stability."

Before long the miners brought their wives and children, the towns were built, and schools sprang up.

A Thousand Years of Gold

Bayard Taylor, a travel writer and poet, came into Central City in 1866 and found what had grown from the seeds Gregory had inadvertently planted.

> [It] is by no means picturesque. The timber has been wholly cut away, except upon some of the more distant steeps, where its dark green is streaked with ghastly marks of fire. The great, awkwardly rounded mountains are cut up and down by the lines of paying 'lodes,' and pitted all over by the holes and heaps of rocks made either by prospectors or to secure claims. Nature seems to be suffering from an attack of confluent small-pox. My experience in California taught me that gold mining utterly ruins the appearance of a country, and therefore I am not surprised at what I see here. On the contrary, this hideous slashing, tearing, and turning upside down is the surest indication of mineral wealth.

Taylor detailed the houses, shops, mills, and saloons in both Central City and neighboring Black Hawk. Not a happy camper by any stretch of the imagination, Taylor complained in his narrative about the high altitude, a bleeding nose, and needing to catch his breath every twenty feet. However, he does come away from the experience impressed with some of the area's early residents.

"In this population of from six to eight thousand souls, one finds representatives of all parts of the United States and Europe. Men of culture and education are plenty, yet not always to be distinguished by their dress or appearance," Taylor wrote. "Society is still agreeably free and unconventional. People are so crowded together, live in so primitive a fashion for the most part, and are, perhaps (many of them), so glad to escape from restraint, that they are more natural, and hence more interesting than in the older States."

Taylor said going on a descent into a mine was one of the necessary things a traveler to the area must endure—and as such agreed to subject himself to the experience.

"It is a moist, unpleasant business," Taylor recounted of his journey into one of the area's larger mines. "As we were returning to the lower drift, there was a sudden smothered bellowing under our feet, the granite heart of the mountain trembled, and our candles were extinguished in an instant. It was not an agreeable sensation, especially when . . . [I was informed] that another blast would follow the first. However, the darkness and uncertainty soon came to an end. We returned to the foot of the ladder, and, after a climb which, in that thin air, was a constant collapse to the lungs, we reached the daylight in a dripping, muddy, and tallow-spotted condition."

When Taylor's tour of the area was over, he embraced the opportunity but reflected on Colorado's gold mining future.

> One thing is certain: the mines of Colorado are among the richest in the world. I doubt whether either California or Nevada contains a greater amount of the precious metals than this section of the Rocky Mountains. These peaks, packed as they are with deep, rich veins seamed and striped with the outcropping of their hidden and reluctantly granted wealth are not yet half explored. They are part of a grand deposit of treasure . . . and if properly worked, will yield a hundred millions a year for a thousand years. Colorado, alone, ought to furnish the amount of the national debt within the next century.

A father-and-son mining team demonstrate double jack drilling.
(Photo by L. McLean, courtesy the Historical Society of Idaho Springs)

Tom's Baby

The Phoenix, the Flag, the Chieftain, the Loch Ness monster. I was surrounded by some of the most famous and unique gold discoveries in Colorado's history. I was in good company: standing next to me in the gold exhibit at the Denver Museum of Nature & Science was geology curator James Hagadorn. Hagadorn explained to me that many of the gold specimens in front of us received their colorful, if mildly unusual, names because of the way they look.

"For instance, we have a piece of gold in our collection called Goldzilla," Hagadorn explained. "If you look at it—it looks just like Godzilla. People see things in gold, in their shape, just like people see things in clouds."

And while all the gold pieces in the museum's collection are breathtaking, I had eyes for only one: Tom's Baby. The massive gold nugget weighs an astonishing ten pounds and takes a position of prominence, resting in its own secure display case. "This piece of gold is relevant to

Coloradans and people from the Rockies because it has such a cool history to it," Hagadorn said. "[And] it is the biggest."

Unlike the other pieces that received their names based on their appearance, Tom's Baby was so named because of the antics of a gold miner more than 129 years ago. Tom Groves, understandably beside himself with excitement after the discovery, eagerly showed it off along the streets of Breckenridge while cradling it in his arms like an infant.

The Biggest

In 1887 miners Tom Groves and Harry Lytton were contracted to work for a mine owner in an area called Farncomb Hill. The two were surprised when on a hot July day they came across an underground pocket, or vug, of gold. Such discoveries were amazingly rare, and the miners removed some 243 ounces of gold from the spot.

Included in that discovery, and at the bottom of the pocket, was the largest gold nugget discovered in the state, then weighing thirteen and a half pounds. According to historian and mining engineer Rick Hague, the two men were afraid the gold would be stolen on their trip back to town so Tom Groves disguised it by wrapping it in a blanket and keeping it under his jacket. But it didn't take long for the news to get out.

"Yesterday hundreds of visitors called on [the assayer] at his office at the concentrator on the west side, to feast their eyes on this find," reported the Breckenridge *Daily Journal.*

The reporter stated that Tom Groves was so excited by the discovery of the nugget and handled it with such care that ". . . the boys declared that it was 'Tom's Baby.' And so it goes." The article went on to say the nugget would later be sent down to Denver so that "Denverites may learn that there are other inducements in Colorado besides Denver town lots."

Like lots of gold. Tom Groves and Harry Lytton were paid a percentage of the gold's worth, and the famous nugget forever left Breckenridge—and for a time disappeared from the pages of history. Hague said the nugget was last seen being handed to the train conductor just before he left the station on his way to Denver.

Lost and Found

At some point, Tom's Baby was procured by Denver's newly started museum, which began in 1900 when Denver residents bought several Colorado collections, including an assortment of gold specimens. According to museum records, Tom's Baby was on display in 1930 before once again disappearing. In 1972 a Breckenridge author began trying to track down the missing gold nugget and was led to vaults in the First Denver National Bank owned by the museum. Tom's Baby was rediscovered there—albeit three pounds lighter and in a box labeled "dinosaur bones." It was concluded that the missing piece had likely broken off in the intervening years. Rediscovered, Tom's Baby was put back on display in the museum in the late '70s.

"This piece is important for its historical aspects," Hagadorn told me as we stared through the protective glass at the specimen. "This piece is important because it is the largest gold nugget in Colorado and it is not necessarily like a nugget that you'd find tumbling down a stream in your pan. If you did, it'd be a very lucky day."

Lucky day indeed. Based on the size and current price of gold at $1,224 per ounce, Tom's Baby would be worth close to $200,000. But its actual value is priceless. Hagadorn said based on the gold nugget's history, uniqueness, and because it is part of the museum's founding collection, narrowing in on a value is almost impossible.

"For us, this collection is closely tied to the museum's deep history," Hagadorn said, adding it was a "mind-blowing" specimen. "It has a value that is not economic; it is historical in nature. These specimens are like art; they are worth whatever anyone is willing to pay for it."

The value and historical significance of Tom's Baby was also recognized in 1887 by the Breckenridge *Daily Journal:* "It will probably be a long time before 'Tom's Baby' will be retired as Colorado's big nugget."

So far, that day still hasn't come.

CHAPTER 3

DIVING FOR GOLD

Clear Creek sat silent and ice-packed just south of Interstate 70 on a cold January morning. The popular stream sees thousands of rafters, anglers, and tourists during the summer, but it was now motionless and clogged with a frosty blue and white winter strata. It's an ice that chokes and strangles the river into stillness at least four months out of the year in the mountains of Colorado.

Even farther from the interstate sits a frontage road that winds along the canyon. Rusted and faded buildings from the area's long-dead mining industry follow the road and litter the river's banks on both sides. Beneath a steep, yellow mine-tailings dump south of Clear Creek squats an ancient shack overlooking the river.

It's impossible to know how old the building is—suffice to say that it's received attention at least once a decade. After the local stone and mortar walls went up, someone installed rough metal sheeting to its front and on the roof. And sometime in the last several years Ken Reid added a wooden sign that reads "Man Cave" above the shack's door. The door is held shut with a massive rusty chain.

Reid is a big man, fifty-three-years-old with a giant black beard, perfect teeth, and a cowboy hat faded to where its color is potentially brownish. He talks in a voice so low, listeners often have to strain to reach its depths. He's quick to share an earthshaking laugh, a bit of advice, or a hard-won lesson—and his blue eyes sparkle perpetually with gold fever. Ken Reid looks like he stepped out of a hundred-year-old sepia-toned photograph of the Old West.

"The Cave," as he calls it, looks out over a large portion of his mineral claim along Clear Creek. He said it's a spot that has earned him quite a bit of money over the years. A port-a-potty leans against one wall and behind it sits a generator, which he tugged on several times. It choked and rumbled to life, sending electricity to a handful of random lights dangling from the ceiling that flickered on and illuminated the building's dark interior. Inside there's a cast-iron stove in one corner that he periodically stoked with wood. The heat was just enough to scare off the worst of the cold, which creeps up from the river and slides down the mountain to meet at the metal shack. Outside it was twenty degrees, but inside it was almost warm, smelling of burning wood, generator exhaust, and cigarette smoke.

Around the room lay broken computer parts for rare mineral salvage, mining equipment, an assortment of chemicals, a beaten, brown leather chair—and a plastic tub resting on two upended buckets. Ken Reid is a full-time, professional gold prospector and one of the last to still earn a living from working the cold water and dirt of Clear Creek.

During the warmer months, he climbs into a well-worn wet suit and dives beneath the river's famously strong currents—currents that, during the summer, often take several lives a year in rafting and other river-related accidents.

The stream collects water from melting snow as far back as the Continental Divide. Most of the year it is bitter cold and strong enough in places to pull a full-grown man off his feet.

It's under that water where Reid sucks dirt and stone into the hose of his underwater dredge. Like a powerful vacuum cleaner, it pulls in rocks and other material, sorts through it, and returns the unwanted portion back to the stream. It is in the stuff left behind in his dredge that Reid finds the gold.

When winter comes blowing down Clear Creek, Reid packs up his equipment and brings his operation back indoors to the Cave. There he pans through some of the finer dirt he collected over the summer. The material rested in large tubs on the floor, located throughout the room. In that dirt was hidden the finest gold dust.

Reid said he tries to get into the water to go diving with his dredge as soon as he can but doesn't dive in water during the winter—at least not anymore. Years ago he discovered a way to inject hot water into his wet suit so that he could go into the frozen waters, under the ice, and continue to operate his equipment. And for the first two days he said it was well worth his time.

"The third day I got out of the water, after about three and a half hours underwater, and a cold front had blown in," Reid said, feeding more wood to the fire. "It was eighteen degrees outside and I had a wet suit flash freeze to my body. I was in trouble."

Ken Reid is full of stories like this. Stories that elicit his conspiratorial laughter, as if you were there with him that unfortunate day, looking in horror at the amalgamation of flesh and wet suit. It turns out in this case, thankfully, he wasn't far from a building with a fireplace and was able to free himself from the frozen suit. It was a mistake he would never make again.

So during the winter he now spends his time panning through the dirt and rock he collected over the warmer months. He poured water into the tub resting on the two buckets and filled a faded green gold pan with dirt and started panning.

No time wasted, Ken Reid is always prospecting. No fleck of the yellow metal is too small to evade his scrutinizing gaze. And he finds it regularly. He says he bought the Cave with the gold he found out in front of it. Looking for gold is what he's done most of his life. And one day he plans to find the mother lode.

Squirrely

Reid said he has prospected for gold for forty-five years but admits he was less than successful during his early attempts. He grew up in Denver and was seven years old when his parents gave him a gold pan to keep him busy while they went fishing in the Rockies.

"I was always out fishing with the family, being out by the river and thinking, *Hey, gold was found in this river and we're fishing next to it. Why can't we find gold today?"* But he was searching for treasure long before then. Arrowheads, antique bottles—as a young kid he was always looking for something. Then he found gold and everything changed.

As a teenager he had a dream that he could go up into the mountains and make money simply by looking for gold. Reid bought an old van and drove up on the weekends and continued searching for the elusive metal.

"A little bit here and there, most of it was fool's gold, a lot of mica," Reid said. "Everything that sparkled was gold. The learning curve was very hard, until you can find somebody to take you under their wing and really show you what you're doing."

Reid went to prospecting stores located near a mall, long ago demolished, called Cinderella City. By talking to a few of the old-timers hanging out in the shops, he said he honed his obsession and gained the skills necessary to make it a reality. He was eighteen years old before he found anything substantial. When it happened he was near the City of Golden just west of Denver. That day he said he found enough gold to get him hooked for life. After that it was an evolution. His thought process turned from simply panning for gold to dredging for it.

To dredge for gold is to use a machine that removes sand and gravel from a streambed. The non-gold material is then sorted out and washed away. Reid now relies on suction dredging, which requires him to be in a wet suit at the bottom of a stream using a hose to pull in the gold-rich rock and dirt.

"How can I move more gravel material? It's a game of volume," Reid said. "I'm beyond the hobby stage. It is more of an obsession and, yes, I do make money at it." He said his passion for finding gold has taken him all over the United States. During his best year he found $62,000 in gold in ninety days.

"I've paid for two pieces of property in gold dust and made my land payments in gold that I mined off the property," he said with a degree of professional pride. Finding gold is hard; finding enough gold to make a living is near impossible. Reid is by all accounts very good at looking for gold and is obviously, and more importantly, good at retrieving it.

One year he came across a stretch of Clear Creek, not far from where George Jackson originally discovered gold on that snowy bank so many years before, which helped to bring about the gold rush. It's the best place for finding gold that he's ever seen, and he's been working that area for the past twenty-five years.

Despite Clear Creek County's history of a gold rush and eighty years of organized mining, Reid believes only 3 to 10 percent of the gold has ever been removed.

"We're sitting on billions of dollars' worth of value here."

But it is hard to imagine where that gold would be, or even if it were possible to remove. Today the majority of the county's nine thousand residents are nestled along a razor-thin valley bookended by the mountains and Interstate 70. Space and tourism are the most important commodities in the county. There's literally no room for new mining operations or, failing that, the essential governmental willingness—especially when tourism dollars glitter ever more brightly than gold.

Those interested in mining or prospecting have to do so along the footsteps and in the ruins of those who have come before. Clear Creek travels sixty-six miles from the Continental Divide to the Great Plains, where it eventually merges with the South Platte. It's along this stretch Reid discovered gold. Like any self-respecting prospector he got a claim, which gives him the right to legally remove the gold from the area, while forbidding all others. Because finding successful pockets of gold has historically been difficult, especially now after nearly one hundred years of mining, Reid said he occasionally finds unfriendly competition.

Every year someone will come down to his portion of the stream and tell him they heard gold was found there. His response is to the point: "'Yeah, but I own the property, I'm the one who found the gold—and I don't want you on that property.' And I have to ask them to leave, sometimes on a daily basis."

He said that over the years people have become "squirrely" about the issue or even hostile. It has gotten bad enough that he sometimes brings his handgun with him.

"I've had some [prospecting] neighbors that were next door that were less than cordial," he said, "and I wouldn't go to my property without a gun."

Gold Fever

Ken Reid said he knows people whom he can trust implicitly with his money. People who will starve before thinking of spending even one dollar of his cash and breaking his trust.

"But they'll fistfight their brother over a flake of gold—and I've seen them do it," Reid said. "Gold fever is a real thing. People see gold and it just boggles their mind to the point where they think that the gold is more valuable than the cash. I'm not going to fistfight you over a flake of gold—but I'm not going to let you take a flake of gold from me either."

Between tools and land, Reid has put $100,000 over the years into looking for the yellow metal. "Run away, do not catch gold fever; it will make you obsessed," he said. "I have spent every bit of gold that I have ever found on acquiring more equipment to go after more gold."

About twenty-five miles away from Ken Reid's Man Cave in the town of Golden, Colorado, gold-panning veteran Bill Chapman leaned against the counter of the prospecting supply business Gold-n-Detectors. Chapman said, with all seriousness, that gold fever is a real condition. One that he's seen in himself and others.

"I have been in the 'hobby' over forty years—and I still have dreams about finding gold and about prospecting," Chapman said. For him, gold fever can be summed up in one word: "lust."

"It is the thrill of the hunt and the thrill of the find," Chapman explained. "And if I get a little gold—that is all well and good." Chapman said going out and finding the precious metal, for almost no money, is a challenge that many are happy to try their hand at.

"But then it gets in your system, it gets in your blood," Chapman said. "And gold fever is a real thing, it is absolutely genuine, and we have seen it amongst our customers." Chapman noted it hits people the worst who go out and are successful at looking for gold the first time.

"There's a lure behind gold that attracts people to it."

Gold Dust Dreams

Back at the Man Cave, Ken Reid was sitting on an overturned white bucket and panning through gravel and dirt with that old green-colored gold pan. Green is used because it's thought that the color of gold stands out more starkly against it.

"Every time I find a piece of gold, I'm the first human being to ever put that into the world market," Reid said. "It's just the allure of how beautiful it is, every piece of gold is different." With a plastic snuffer bottle Reid sucked up the small gold flecks he regularly discovers.

He rarely scores an actual gold nugget and most of the gold he collects is in gold dust. He said oftentimes a nugget, however, is worth more than its weight in gold because of its uniqueness.

"When you find a nice specimen, every one is unique, every one is different, and no two nuggets are the same," Reid said.

It's not unusual to find Reid walking the streets of Idaho Springs during the summer and digging into his deep pockets to pull out a nugget to show off in the bright mountain light.

"Everybody is looking for that big piece of gold; you're more likely to find a five-carat diamond in the Earth's crust than you are a one-ounce gold nugget." With a dreamy look he recalled the story of how he once found a nugget that was shaped exactly like a horse's head. For years it replaced the knight on his chess table.

"I never did find the entire chess set."

Reid said he doesn't have any difficulty finding people willing to take gold instead of cash.

"I've paid for land with gold, I've paid for mining equipment in gold, I've paid for cars in gold," Reid said. Another time he found a piece of wire gold shaped like a spinal column. He traded it to a chiropractor for work on his ailing back. Being hunched over a stream with a gold pan is physically difficult work. Reid said he is able find relief from his aches and pains diving under the stream for hours at a time to look for gold.

"It is still a lot of work digging underwater but I have the buoyancy of the water and it really helps the joints, the back, when I'm laid out in the stream," Reid explained.

"Rock Wrestling"

Under the fast-moving brown water Reid does what he calls "rock wrestling," or digging up and moving boulders to get his dredge's hose under them. He's dove as deep as thirty feet, trying to get at where the gold might be hiding. "I am diving in hypothermic swift water at altitude; these are all dangerous things."

It's not always gold that Reid finds under the water. Over the years he's discovered railroad spikes, railroad tracks, old shovels, broken glass, and other relics from the area's mining history.

"The top five to six feet of the river is man-made trash," Reid said. "Clear Creek is literally filled in over the last 150 years of mining up here with man-made debris. Once you get below that six-foot level you're down to a level where man hasn't been."

History states that some hard rock mines were built under Clear Creek, but Reid said he isn't worried about coming across an old shaft and being sucked into one. He said any shaft that has already come that close to the river is likely already filled in with water, and added he occasionally comes across placer mine shafts in the river that once were located beside it. One in particular he knew was in his area and spent three years looking for with an old photograph.

"I have a 1902 USGS report showing that shaft is sixty feet deep. I want to dive that shaft," Reid said. Because he is considered a prospector, Reid is allowed to remove seventy tons of material a year without a mining permit from the state of Colorado.

"I don't count how many tons of gravel that I remove—but the year I find seventy tons of gold I'll get a permit," he said, shaking the Man Cave with his laughter.

Reid has supplemented his income with gold prospecting but this year he plans to make his entire living looking for and finding gold. Formerly the operator of an antique and pawn store in Idaho Springs, his business wasn't making a profit and he had to close its doors.

"I'm going to dedicate this entire year, this is what I'm going to do," Reid said. "I'm working a part-time job to get through the winter, but come springtime I'm going digging."

Reid said that if he ever does come across a lot of gold, he would spend the money traveling the world and looking for more gold until he was broke again.

"This is no get-rich-quick scheme." But for now, while the winter winds still tugged at the outside of his Man Cave and the drifts of snow collected outside the building, he'll spend his time panning through buckets of dirt for more gold dust. Gold dust is what Reid finds the most of, adding that 90 percent of his gold weight is found in dust and little flakes.

"You'll find pounds of that gold dust before you find a nugget," he said. "But nuggets take a premium. You can get four or five times on a very characteristic piece of nugget gold."

Removing and collecting such small pieces of gold requires tremendous patience. It's a difficult pastime and he said people often don't, or can't, keep at it.

"A lot of people don't last at it because it is so tedious," he explained. "They want to run out and get all the big pieces of gold and think they're going to go off skipping to the bank and get rich, rich, rich."

He admitted some have stumbled across a big piece of gold or two—but often it's the small stuff that provides a decent and regular payout. "Why would you want to throw away gold when it is here for the taking?"

He uses a dredge for most of his major gold operation during the summer, but he's not ashamed to go back to using a gold pan.

"It all starts with a pan. I don't care where you're prospecting at—you're not going to go in with a million-dollar track hoe and dig up gravel and say, 'OK, there's the gold,'" Reid said. "For anybody who's prospecting, 90 percent of the people are using the pan."

The idea is a beginner starts off with an affordable gold pan costing maybe $12 and then moves up in equipment as they find more gold.

"I started with a pan when I was a little kid and every year I just progressively added on," Reid said, pointing to the various corners of the Man Cave. "Here we've got a screen over here, we've got a power crusher there, we've got a kiln there, we've got a magnetic separator there, we've got a vented hood for our assessor lab stuffed over here. There's a shaker table and a power screener outside."

But watching Reid use a gold pan is like watching someone who has mastered their craft. His technique is fast and efficient. The way he used the water to pull the rock and dirt from the pan is second nature. Soon all that was left behind was the fine, heavier magnetic material and gold.

It didn't take him long before ultrabright flecks of gold started appearing amid the fine black sand consisting largely of iron and magnetite. He took a large magnet and moved it around inside the pan, pulling the black material out. Before long only a small line of yellow dust remained. He sucked up the gold dust and started again with another scoop of gravel. He submerged the pan in water, shook it, and let a wavelike motion of water remove the dirt.

He reached in with his massive fingers and removed the stones, tossing them aside. "Just like any trade, anybody can go out and buy a saw and a hammer and call themselves a carpenter, but it takes years and years of work to become a master carpenter."

Reid said it is the same with gold panning. Again he reduced the rock and dirt to black sand in which several small flakes of yellow gold peered up at him through the dim and smoky light. "I find gold every day I look for it."

Despite years of gold mining and prospecting, he stated all the gold hasn't been removed from Clear Creek. Rather, erosion helps to replenish it every year.

"The gravel bed is always moving. If you actually look at the gravel bed, it is a flowing mass; it's moving at glacial speed," Reid said.

Helping Hand

Reid said he's happy to share the stream with the summer recreational users. He added many times the rafting guides will see where he is in the water and try to steer around him. And in return he's already in the water if someone falls out of a boat. He's been in the right place and at the right time to help drag people from the stream.

"I've had people puking water on me if they've been in the water long enough. I've had people airlifted out of this valley that I've helped extricate out of this water. One day I took four people out."

Rafters are not the only people using the county's waterways. Reid said once the economy starts to suffer, people begin to look at the hills again and dream of gold.

"You get the rich investor who thinks he's going to invest in a mine and he's going to become richer. Then you've got the guy who can't barely afford a sluice box, pan, or gas to get here. He comes out and he's starving to death on the side of the river. If he stays off the goddamn bar stool and works hard—he can make good wages over the summer."

People down on their luck have gone to Colorado's streams looking for a second chance before Colorado was admitted to the Union in 1876 as the thirty-eighth state. During the 1930s a public works program was created in Denver to pair seasoned miners and prospectors with people who lost their jobs to the Great Depression. The program was profitable and gold panners claimed $1 per day from the South Platte River, which Clear Creek feeds into. It's thought that the stream's placer gold deposits weren't cleaned out during the gold rush but by people trying to make a living during those lean years.

According to Reid, every year people come to him and ask where to go to find gold, and what it is they need to do to get it.

"There's very few people who stick with it. This is the real gold fever—when you sit here and spend years chasing it."

When Reid is working beneath the river, a six-hour day often feels double that.

"If you don't like digging ditches at the surface, you're not going to like it underwater much better," he said. "If you look at your real successful treasure hunters, if you look at your real successful miners—they're the people with perseverance."

The Life Style

Over the years of rock wrestling and diving under the freezing waters Reid said he has "good days" and "really good days."

"I've had days where it's just been nothing but a fight all day long. Plug ups and rock jams, cave-ins, but you gotta do it," Reid said. "But just like any job you have your good days and your bad days too. Good days in this job you can get rich fast."

Admitting to spending every ounce of gold on trying to find more gold, Reid isn't wealthy yet. He's also not getting any younger and the work doesn't get any easier, but he said he continues to descend into the water every year and put his safety at risk because it still gives him an adrenaline rush.

"You always want more and that's the fever of it. Kind of like the successful businessman who wants to keep making more money. You just never know. Come springtime when I fire up that dredge and get back in that water and punch another hole—it could be the year that I don't have to do it no more."

He likes working outside and looking for gold, and not being stuck at a desk and working a nine-to-five job. Reid said he has lived a lifestyle his father and uncles once dreamed of around their summer vacation campfires.

"I've done well. I probably haven't been the most productive person in American society, but I've chased my dreams."

And he's not alone.

CHAPTER 4

ALADDIN'S CAVE

Sunglasses, neon hard hat, vest, hoodie—Brad Poulson was wrapped in layers of clothing specifically designed to alert someone else to his presence. Safety in the world we were about to enter is key. I waited as he walked around a company SUV with the words "The Cripple Creek & Victor Gold Mine" printed on the side. He checked the mirrors and tires, put a magnetic antenna on the top, and did a careful check around the vehicle.

We were about to drive into the last commercial gold mine in Colorado, which is owned by the Newmont Mining Corporation—the second-largest gold mining company in the world. Newmont purchased the surface mine from its previous owners in 2015 for $820 million and added the operation to its roster of others located in New Zealand, Australia, Indonesia, Suriname, Peru, and Nevada.

Poulson took his time. Every action, every word he uses is specific and calculated; he clearly knows what he is doing and has done it a thousand times before. The company's communications specialist met me in the historic mining town of Cripple Creek. Obviously knowledgeable, he speaks succinctly and clearly, choosing his words carefully and often stressing the last syllable of the last word in a sentence. When he finished his safety checks we climbed into the vehicle, he beeped the horn several times to let anyone nearby know he was backing up, and then began the short drive to the mine.

With its own traffic laws, vehicles the size of houses, and a man-made ashen gray geography, it felt like we were about to drive across the surface of another planet. Poulson, who has worked for the mine for the past three years, told me that Newmont employs 580 people at its Colorado operation, paying them on average $79,000 a year with benefits. For various reasons, including the fact that many of the homes in Victor and its sister city, Cripple Creek, are historic, most of the mine's employees live outside the old gold towns in the northern part of Teller County.

A vehicle gate let us in, and Poulson used his radio to ask for permission to enter. The sky was blue, but almost all the recognizable landmarks of the surrounding mountains were hidden behind hills and berms—conversely, nearly all the mine's workings are hidden from the outside. We heard the thumping of a distant rock crusher just under the crisp radio chatter prepping for an explosive detonation. Poulson received permission and pulled onto the mine's property.

"Here at CC&V because of the way that this area developed, the property is actually privately owned," Poulson said. "To begin with, this was a ranching area, this was high summer pastures for ranches along the Front Range. Some of the land was sold off by the federal government to ranchers, and then when gold was discovered the federal government sold off the land in patented mineral claims and those patented mineral claims were aggregated over time to the land package that CC&V currently owns or leases."

The discovery of gold in this area led to a mining claim purchasing rush in 1891. At one time the area had some five hundred different mines that were eventually purchased and aggregated. It's been estimated that there are some 2,500 miles of underground mine workings that exist between the towns of Cripple Creek and Victor. In fact, the sixty-five ounces of gold covering the state's capitol dome in downtown Denver originally came from the mining area in 1908 in honor of the gold rush. The Cripple Creek & Victor Gold Mine most recently provided the gold in 2013 when the dome needed to be replated. As a schoolkid in Colorado I remember tours of the capitol dome and the tale of at least one senator

A Caterpillar mine haul truck is loaded with gold ore at the Cripple Creek & Victor Gold Mine. *(Courtesy of CC&V)*

who snuck up to the dome after every rainstorm to collect an untold amount of gold dust.

Located southwest of Pikes Peak, the CC&V, a surface gold mine, was started in 1976 shortly after the deregulation of gold by the federal government. Today the mine's property stretches nearly six thousand acres and quite literally dwarfs the nearby towns.

"In the 1970s the price of gold was deregulated, and mining came back to the district, eventually building to the regulated, large-scale operation we have now," Poulson said. The mine currently has twenty-five Caterpillar mine haul trucks to move the ore around. Having never seen one in person, I'm stunned by their massive size. Costing about $5 million and capable of carrying 250 tons—or more than the weight of the Statue of Liberty—the vehicles are enormous. A person essentially has to climb a long metal staircase to get to the driver's cabin at the top. Poulson explained for every truck carrying 250 tons, roughly six ounces of gold are recovered. An ounce of gold is roughly the size of the top third of a pinky finger.

“They’re really giant computers on wheels that are being monitored via satellite back to our dispatch center,” Poulson said, as I stared openmouthed as one passed by in the other lane. So large in fact that there was a good chance the driver wasn’t even aware that we were on the road with him. Poulson added the truck’s engine temperature, hydraulic pressure, speed, location, and what they’re hauling were all monitored.

“The thing is, the trucks are twenty-seven feet wide, twenty-five feet tall, and about forty feet long,” Poulson said. “So it’s like driving a two-story house. When they’re in the operator’s cab on the left-hand side of the truck, they literally can’t see the right-hand side of the road for 120 feet. So they drive on the left-hand side of the road so that they can see the berm.”

Everywhere you go in a modern surface mining operation, you’ll see berms at least half as tall as the tallest tire on the road. In this case, the tires are twelve feet tall, so the berms are at least six feet tall. If a driver loses control of a truck, or if there’s a brake failure, the berm is engineered to stop the vehicle from going over the side of one of the deep surface mines. Those mines in fact dwarf the trucks, sometimes going down over a thousand feet in what looks like an inverted pyramid. The size of it can easily boggle the mind of the uninitiated.

The Wild Horse

Poulson and I got out of the truck at the bottom of the six-hundred-foot-deep Wild Horse extension surface mine. I couldn’t help but feel like I was in one of the moon’s craters. These types of mines consist of “walls and benches” with walls between thirty-five feet and seventy feet tall and twenty-foot benches, used to stop tumbling rock. The whole thing looks like stairs for a giant. Poulson explained that the process starts in part with fifty-foot drills that are set up in GPS-specified locations. Approximately 250 of the holes are drilled in one area about fifteen feet apart. A detonator the size of a man’s fist is put into the bottom and the hole is pumped full of ammonium nitrate and diesel fuel.

“We pump that full of explosives except for the top ten feet . . . I am talking about people who have mega skills,” Poulson said. “I am talking about people who are skilled in what they do and handling every aspect of this."

The company's blast technicians then backfill the new hole with crushed rock so that the explosives go off sideways and fracture the earth. In doing the surface mining, up to 1,100 holes can be drilled and detonated, but no more.

"We have certain regulated permit caps on the amount of seismic motion that we can create within the earth," Poulson said. "We want to keep it well below that, so we're in permit—but most importantly so that we minimize the seismic impact to our neighbors."

The explosion will turn the rock into rubble that is five feet in size or less, which is what is needed to fit into the mine's crusher. The rubble, or "shot muck," is shoveled into the back of the mining trucks. A digital readout on the side tells how many tons they're carrying. Poulson said the area's original miners were digging out gold from veins or removing high-grade ore. What the Cripple Creek & Victor Gold Mine is hauling to their crusher is disseminated ore. It takes many tons of disseminated ore to get an ounce of gold.

"That's why we have to move so much of it to make it economic," Poulson said. Every year sixty million tons of rock is moved, which includes twenty million tons of ore, and forty million tons of overburden, or waste rock.

A Vug of Gold

We got back in the company truck and headed out of the Wild Horse extension surface mine to see one of the other, larger mines. Poulson slowed down as we pulled in behind one of the massive haul trucks.

"You never pass a haul truck without permission and if it is moving you don't even ask for permission, you just get in line," he said. "Haul trucks have the right of way." It's easy to see why. Anything the size of a multistory house with wheels that belong on a monster truck is going to win a game of chicken every time. Poulson told me the area was historically so rich in gold that the miners could dig out large veins of gold, sometimes as big as five feet wide. In 1914, a miner discovered a "vug," or giant pocket, of gold in the area. It yielded sixty thousand ounces of gold.

The Cripple Creek & Victor Gold Mine's Cresson Surface Mine. *(Courtesy of CC&V)*

"The vug of gold disclosed the beauty of an Aladdin's cave and the wealth of the United States Mint," reported a 1918 edition of *The Mining American*. It wasn't the last vug of gold discovered in the area.

"Thirty-eight coming up the Joe Dandy, in the wrong way, and going around the blade," the radio chirped incomprehensibly, taking me from thoughts of massive caves filled with gold. Poulson said the mine's exploration geologists work to create three-dimensional underground images to get an idea of what lies beneath the surface.

"An engineer can apply very sophisticated tools to determine the cost inputs, versus the revenue inputs, from the ore body that is being mined," Poulson explained. "Ultimately that determines the shape of the surface mine."

The Cripple Creek & Victor Gold Mine currently has two surface mines in operation and a third and fourth in the process of being opened. The 1,200-foot-deep Cresson surface mine shows scars along its steps of the 100-year-old mines that came before. Today's miners often come across and have to backfill those original mine shafts and tunnels.

"The old-timers went much deeper than where we are," Poulson said.

A LIDAR unit, which stands for Light Detection and Ranging, uses lasers like a radar uses sound. The mine uses the lasers and a slope radar to measure and draw a picture of the mine's wall to help determine if there's going to be a collapse. There are also old mine stopes, or underground empty spaces where ore was once extracted, large enough to swallow one of the big trucks. Poulson said the mine is constantly monitoring and keeping track of where historic mining activity took place and where voids might appear.

"We have historic mapping, that is now computerized . . . and then all of our exploration and blast hole drilling is logged with GPSs and they find voids," Poulson explained. When voids are found they are logged, mapped, blasted, and filled with rock by remote controlled vehicles to keep the miners out of danger.

"We're not bragging—but modern mining is safe because we maintain a safety focus," Poulson said.

"Pregnant with Gold and Silver"

The ore is transported to a crusher for processing. Crushers grind the stone like a massive mortar and pestle. The mine's primary crusher works through seventy thousand tons of ore a day, and twenty million tons a year. The ore then goes past an electromagnet that pulls out old lunch buckets, ore tracks, lamps, or anything metallic that doesn't belong. It then goes on a conveyor belt where it is sorted into different sizes or sent back to another crusher. There are a series of conveyor belt tracks that look like the highest, most unpleasant roller coaster in the world.

Lime is mixed in with the crushed ore to keep from evaporating the process solution's sodium cyanide, used to leach the gold and silver out of the ore. The crushed ore is stacked in these gray hills, and the solution is dripped out onto them. Since the early '90s, four hundred million tons of crushed ore have been collected in this part of the property. The gold and silver dissolves in the diluted mixture of sodium cyanide in water and flows down with the solution to self-contained underground ponds or wells. Poulson said none of the fluid escapes this double- and in some areas triple-lined facility, and the solution is recycled back through the process after the gold and silver are removed.

"The gold and silver is more attracted to the diluted mixture of sodium cyanide in the water than it is to the elements in the rock," he said. "So the gold and silver leave the ore and attaches to the sodium cyanide and the process solution becomes pregnant with gold and silver."

That gold- and silver-rich solution is then pumped to another facility where the minerals are removed from the solution and put into a furnace. The melted metals are poured into "button" molds that weigh about sixty pounds and consists of about 63 percent gold, 30 percent silver, and some impurities ranging from copper to iron. It is then sent to another facility out of state where the metals are further refined.

"Last year we produced about 190,000 ounces of gold and about 60,000 ounces of silver," Poulson said, driving alongside the leaching pads. "Next year we're estimating over 350,000 ounces of gold because of our investment in new process facilities."

That gold is sold into the market. Gold can be used for bullion, jewelry, or a reserve currency for governments; as an investment; and in the high-tech manufacture of smartphones, avionics, and satellites.

Builders of high-rises use gold in glass to reinforce it and to filter out radiation. Colorado's only commercial gold mining operation is growing and has plans for expanding and may one day even consider plans for more traditional underground mining operations.

The gold will eventually become depleted from the area. This inevitability comes to all mines and across all eras. Strict regulations on modern mining will also require the company to reclaim the area and match the surrounding geography and ecology.

But the search will continue as it has since antiquity and the Cripple Creek & Victor Gold Mine is not alone in looking for gold in Colorado.

Across the state, and in the historic mining districts, there are people picking up gold pans and prospecting supplies for the first time. And then there are those who have hunted for gold their whole lives.

CHAPTER 5

CHASING A BULLET

The days when men looked for their fortunes in the cold streams of Colorado's mountains have faded like an old photograph but never fully died away. In 2011 with the price of gold reaching almost $2,000 an ounce, a staggering amount, the interest in searching for the rare metal reached new heights. Modern prospectors are breathing new life into mining organizations such as the Gold Prospectors of the Rockies and the colorfully named Colorado Chapter of the International Order of Ragged Ass Miners.

The steady stream of fortune hunters visiting a store located two and a half blocks north of Clear Creek in the City of Golden is also a good indication of the metal's ever-powerful lure.

The moderately sized shop is stuffed full of books, equipment, tools, and other items needed in the quest for gold. It is also the only store in the state that still caters solely to modern-day prospectors. Bill Chapman started Gold-n-Detectors in 1995 after he retired from the Jefferson County Sheriff's Office. Several years ago, he sold the business to Louise Smyth, but he still shows up most days to answer phones, talk to prospectors, and tell stories as rich as the veins mined 150 years ago. Chapman wore a black vest and enthusiastically sipped from a giant plastic coffee mug—the kind you might find in a convenience store on a

Veteran gold prospector Bill Chapman still offers advice at Gold-n-Detectors in the town of Golden. *(Photo by Chancey Bush)*

road trip through Kansas. The cup looked like it could hold enough caffeine to drop a rhino.

Grandfatherly and with the air of a retired law enforcement officer, Chapman is happy to give advice to would-be prospectors. However, he only gives it once, and if someone doesn't follow it, he won't give it again. When he said this, it didn't sound so much like a threat as it did a promise—a promise he has clearly kept more than once. Over the past twenty years, Chapman has watched interest in gold prospecting grow. It has gotten to the point where he sees hundreds of new prosecutors open the front door to the prospecting shop every year. It is also well-known in the prospecting community that Gold-n-Detectors buys gold. Not the kind found on old watches or in the many necklaces of Mr. T—Chapman's store buys gold dug from the earth.

He showed me the vials of gold dust or nuggets purchased from people fresh from a day out on the stream. Chapman said people come to him to sell their findings, which he's happy to buy, based on the current value of gold that day. Unlike tennis or fishing, he said this is an activity that pays off.

"We sometimes have guys come in with more than we can afford to buy," Chapman said, eyeing one vial through a pair of broad prescription glasses.

The lure of gold is a strong one. The night before I met Bill Chapman, he had a dream. He said he was sitting by the side of a river with a gold pan looking for gold. It's not something he does very often anymore as he wades into the later years of his retirement. But for more than four decades, Chapman has dreamt of gold and over the last twenty years has helped others to share his dream. His journey to become a prospector, and ultimately to start the most successful prospecting supply business in the state, began with a murder and a lost bullet.

A Bird's Nest

Chapman will gladly show anyone his biggest gold find ever. He still has it and, despite several offers, has no plans to sell. He walked across the store, past the shelves stuffed with prospecting tools and metal detectors, to a narrow glass cabinet tucked away to one side of the cashier. He reached inside and pulled out a clear plastic box. Inside sat a giant chunk of gold. He said the specimen is an example of "bird's nest wire gold"—and that name fits it perfectly. Golden wire and threads branch out from a core of gold in a chaotic but beautiful sprawl. The metal was removed from the rock by disintegrating the rock with chemicals but leaving the gold intact.

"It is a unique form of gold that is only found in the rocks. It is not found liberated in nature; it is too fragile, and it is just a mess—it looks like a Brillo pad. And it will fall apart if it is exposed."

Chapman said he's never been tempted to sell it because he hasn't found a larger piece of gold yet. The piece wasn't found in a stream but with a metal detector. Though Chapman certainly knows his gold panning, sluicing, and high-banking methods of gold hunting, it was metal detecting that originally got him hooked.

"Work Like Idiots"

In 1964, his father found his lost Eagle Scout ring with a World War II–era mine detector at a scouting camporee. They also found coins, scout badges, and a knife—real treasure for a young boy and he was hooked.

In 1971, Chapman began hunting for lost treasure with a metal detector, after building one of his own. It wasn't long after that he moved to Colorado, got into law enforcement, and learned the nuances of hunting for gold. He and another deputy would spend their weekends looking for the metal in the mountains above Denver.

"We had a great time; we would just put a six-pack in the creek, to keep it cold, have sandwiches, and dig and work like idiots because we were young and could do that," Chapman said. "But it was fun, we had a good time, we found some gold, we met some other people who did the hobby, and afterward I got married. That kind of killed our partnership because I didn't have weekends to go out anymore. I've been in the hobby for a long, long time, and I learned the way most of us learn: the school of hard knocks."

One such hard-won lesson came after an afternoon of working along Clear Creek with his prospecting partner. A man came up to them and asked if he could place his sluice box downstream from where they were working.

"Half a mile in either direction there was nobody on the creek," Chapman recalled. "'Well, hell no, it is a free creek.' We thought he meant downstream, but he meant within a foot of where we were working."

The two continued to collaborate to find gold in their sluice box and said they had found some while the man mostly sat on a stool and smoked cigarettes.

"At the end of that day he said, 'Now, I saw you boys didn't know what you were doing, so this is a lesson.' He had more gold in his box than we had in ours and it was gold that we had worked for," Chapman said. "And he says, 'Here's what you were doing wrong.' From the roadside, he could see we were making mistakes, and there is no such thing as a free education. That is a lesson that I will remember forever."

Chapman and his partner were accidentally pushing the gold out of their sluice box downstream.

"He saw us doing it, and he knew it was wrong, and it was a lesson that we learned and a mistake we didn't duplicate," Chapman said. "But that was a life's lesson for us. Anyway, we were always getting gold but once we learned that lesson we got more gold."

Stories like this one aren't uncommon among Denver's gold

prospecting royalty. Many of those who have been at the game of hunting for gold have a greenhorn story that they can proudly recount. The stories serve as both a reminder of how far the storyteller has come through the school of hard knocks and as an Aesop-like lesson without talking animals. The tales are also as familiar as condos at ski resorts—but their telling does in no way diminish the lesson: looking for gold is hard and can rarely be accomplished without a mentor.

Crime Scene

The time came when Chapman was using his metal detecting skills to help find evidence at crime scenes.

"For years, I'd been dragging my metal detector to crime scenes because it was a much better tool . . . the metal detector that we had was so poor it was virtually useless," Chapman said. "For years, I had petitioned for a new metal detector for the lab and was told no."

Then in 1995, there was an incident that saw two people murdered in a food store. The gunman then went to his van and took out a high-powered rifle and waited for officers to respond. When the first deputy arrived, the man opened fire from across the parking lot and killed him by shooting into his patrol car. The man was caught and later sentenced to life in prison.

Chapman was a deputy sheriff criminalist at the time, and it was his job to find the ballistic evidence. There was one bullet he couldn't find; he decided he needed better equipment. Circumstances had changed at his department, and he was given permission to buy a new metal detector. He said there was only one metal detector store in town.

"When I called them I was surprised to learn that they'd gone out of business," Chapman said. He then called the manufacturer and told them he was getting ready to retire from the sheriff's office and he could open a dealership for them.

"It had been a hobby of mine, both a prospecting hobby and metal detecting hobby, for decades. And since I was due to retire, my wife told me I better find something to do to occupy my time," Chapman said. "So I did."

But he never did find that missing bullet.

"I've found bullet holes in houses, but it is just an enormous task to look for a small bullet that goes many hundreds or thousands of yards," Chapman explained. "And all the land that can be in between where it originated and where it finally ends up. No, I did not. I did not find that original bullet."

Treasure Hunters

For a long time, Chapman said he had trouble finding the right places to hunt for gold. Looking for gold in streams is best done by hunting for it in places where the original prospectors did during the gold rush, but finding larger portions of gold with a metal detector is a closely guarded secret. He later learned that many successful metal detectors were finding gold in the tailing piles or dumps of old gold mines.

"We started hunting the dumps in Leadville; obscene amounts of gold came out of those dumps, I mean we're talking one piece that was $30,000. For a hobby that ain't bad."

Chapman said the gold wasn't taken out by the original miners because they overlooked it.

"They'd look at it and say, 'There's no gold there,' and they toss it out in the dump pile. In fact, there was gold there and the metal detector could see it where the eye could not."

While looking for gold with metal detectors brought in larger and more wealthy finds for metal detectors, there was a drawback. Over time the more gold they found, the less there was for future generations to come across.

"The gold that we find is not being replenished in the metal detecting arena," Chapman said. "Every year the creeks replenish more gold as it washes into the creek and spring thaw and runoff. But not in the tailing piles. Once that piece of gold is found and out of the ground, it's gone and won't be back, and nothing will show up to replace it."

Also, many of the areas ripe for discovering gold with metal detectors are rapidly dwindling, places Chapman and his fellow gold hunters once poured over in the mountains have since become neighborhoods at ski resorts.

As the locations for gold have become scarce and the interest in finding it increase, Chapman said the competition can become severe and even hostile.

"A lot of people like to travel to the Arkansas River Drainage, Cache Creek, in particular, that's just been plundered by weekend prospectors," he said. "I hear some are very protective of their areas. They get downright nasty."

Chapman said two years ago a man from Texas was in that area looking for gold with a sidearm.

"He was threatening people, and it's part of the national forest, so the parks people had to be called to deal with this guy," Chapman said. "That's not why we're there; we're there to have fun—and that's not fun when it becomes that serious."

There's also a man he refers to as "Crazy Chuck" who regularly carries a gun with him while out looking for gold.

"And we call him that for a reason: he packs heat and he will pull a gun on you if he catches you on his claim. And for a while, he had a night watchman up there to keep people off of it."

Chapman hasn't always operated the only metal detecting and prospecting store in Colorado. Over the years, several have come and gone.

"When gold went up to its record high four years ago we couldn't keep stuff in stock fast enough. It was a mini gold rush, and it was exciting."

He said while it doesn't happen often, people do, however, strike it rich. He said two prospectors used to come into his store with briefcases full of gold they had discovered.

"They had found tens of thousands of dollars, maybe even up to $100,000, in these briefcases," Chapman recalled, his eyes lost for a moment in the memory. "They would open these briefcases on the counter up here, and people would just look in—talk about gold fever."

Those who do well in looking for and finding gold are the ones who are patient. He advised it takes lots of perseverance to be successful in the hunt for gold. Chapman said there was just something special about being able to get a gold pan and go to a creek and find gold like the old-timers did.

"You look at the color in the pan and go, 'Holy cow, I did this myself; this is gold.' It's exciting, and that excitement reinvents itself every time you go out to the creek."

Newcomers sometimes go out into the mountains, with supplies from his shop, without listening to his advice, and find nothing but heartache and a sore back.

"So, 'It's all a humbug, there isn't any gold left,' and that's totally untrue. There's lots of gold to be found yet."

CHAPTER 6

HARD ROCK MINER

Not far from the cold banks of Clear Creek sits the skeletal remains of the Stanley. The mine's largest building dominates the south bank of the river like a dinosaur, just one of many that did not survive the industry's inevitable demise. Rusted red and yellow, its windows are now empty black holes, and its full name, "The Stanley Mines CO," is still legible in bold, dark letters on its face.

For more than a century the iconic building has stood guard over its considerable wealth. Today it silently watches the seasonal traffic that winds its way up and down the interstate.

At one time the gold mine was one of the largest producing operations in Colorado. According to local legend, the mines were dug not far from where early prospectors discovered the remnants of much older Spanish mining operations that predated the gold rush by more than one hundred years.

A Labyrinth

In 1864, once miners had depleted the obvious gold in the placer deposits along the stream, they turned their eyes to the mountains to begin looking for the source. An influx of both Cornish miners from England and

The Stanley, one of the largest gold-producing mining operations in Colorado. *(Photo by Chancey Bush)*

copious amounts of Civil War cannon powder not used in the conflict helped make this possible. The powder was an obvious necessity for getting into the rock, but the Cornish also brought a wealth of experience. The miners of Cornwall had toiled in the earth for precious metals since the time of the Romans.

Early on the Stanley went down three levels and produced nearly two tons of ore a day—ore that provided nine to ten ounces of gold. In 1879 John M. Dumont sold the mining claims to a California group that increased its development. The mine was sold again to a group from England in 1880. At its peak the Stanley's tunnels went down seven levels, and local tales have it that the donkeys used in the mine's operation didn't come out again until dead of either old age or overwork.

Coming into the area by train in 1894, professor and journalist Arthur Lakes wanted to tour and write about a typical gold mine of the era for the readers of *The Colliery Engineer and Metal Miner.*

"We found just what we wanted in the Stanley Consolidated Mine at Idaho Springs. . . . Passing through the town about half a mile up into the canyon, we come across the mills, buildings, plant, etc. of the extensive property of the consolidated Stanley situated on the banks of the stream."

In both directions along the canyon and to the very top of the mountain he saw old and new mining dumps, tunnels, and buildings associated with the mine. The area was riddled with signs of mining since its feverish inception in 1859.

Lakes noticed giant, deep fissures in the ground on the north side of the stream where an ore vein was picked clean. He also noted the surrounding hillsides, once heavily vegetated, were completely deforested to provide timbering for the ever-expanding mines. An illustration depicting a cross section of the Stanley Mines shows a deep, sprawling labyrinth connected on both sides of the river and, at that time, six levels beneath it.

Lakes would have known that a mine tunnel is crafted by blood, sweat, dynamite, and drills. Veering into the darkness, they are created by men chasing an elusive mineral deposit that's fated to die away. From the mine's yawning entrance, the tunnels stretch out before it sometimes for miles. A fine talcum-like dust or mud collects on the ground between the mine cart railroad ties. The tracks, used for ore carts, extend into the tunnels with a steady, rhythmic tattoo.

The rock walls, sweating in some places, bear the marks of the hunt for gold. Further inside, with a lamp punctuating the absolute, cold darkness, the smell of soil and rock fill the air. It's a deep odor, tinged with the acidity of crushed stone and earth. What can't be smelled, only experienced, is the thin air. It causes the lungs to work a little harder. Miners didn't smoke inside the mines. It's generally considered a bad idea to cause anything to get in the way of much-needed oxygen. There's the sense of the countless tons of rock continually pressing down on the straining timbers, rock that is forever leaning into the empty space below it. Under the low ceiling, a tingling sense of claustrophobia dances around the subconscious.

Lakes entered the first level on the surface and noted that the mine was "well and carefully timbered for the first few hundred feet, owing to the nearness of the surface making the roof and walls of rock loose and unsafe." In one portion of the mine, Lakes spotted that the stulls, or upright timbers, were straining under the incredible weight.

"These stulls in this mine, with inclined walls, have to be very thick and strong to support the enormous pressure. . . . So powerful was this

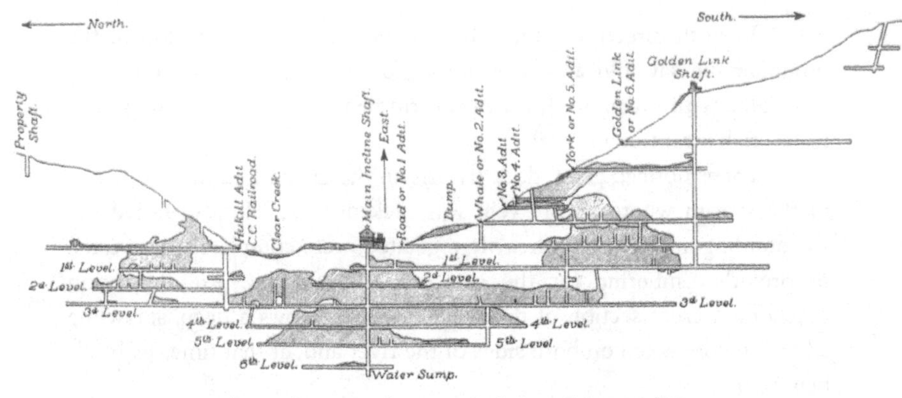

A cross-section view of the Stanley Consolidated Mine. It is said donkeys used in the mining operation didn't come out again until dead of old age or overwork. *(from* The Colliery—Engineer Metal Mining, *1894)*

pressure that I saw cross-stulls two feet thick and only five feet long; split, broken, bent and crushed."

One story goes that the miners working the tunnels almost broke into a deep shaft under Clear Creek. The hole was filled with a wooden plug after water began pouring out. Given the danger of drilling under a river the miners soon abandoned the tunnel.

By the time a United States Geological Service report was published in 1908, the mines consisted of more than four miles of tunnels. In 1910 mining activity faltered and the major mining operation finally stopped due to what was likely mismanagement and a lack of new mine development.

The Stanley was opened and last operated in 1942 before President Franklin Roosevelt executed Order L-208, which stopped the country's mining of precious minerals. Miners were instead to focus their subterranean efforts on mining for metals needed strictly for fighting World War II. It was a crippling and ultimately fatal blow to many of Colorado's mining communities and notably Clear Creek County. Mines were abandoned and mills grew silent for the first time since the gold rush.

The lure of gold is tenacious. Over the years small mining operations, mostly consisting of just a few men, began to operate in the county once again. The new mines were reduced in scope, but the danger never diminished.

A Rescue Attempt

Thirty-five years after Roosevelt's decree, with the price of gold and silver on the rise, forty-eight-year-old Cliff Morrison and three others began working to get the Stanley Mines open once again. The dirty and difficult labor was taken in stride—the four were experienced miners and knew their business.

Morrison's father was a silent film actor and a nearby town was named after his family. Having grown up in the county, Morrison knew the area well and was known for his wry sense of humor and enthusiasm. Morrison was working inside the mine to shore up rotting timbers on May 26, 1977.

The story reported in the *Clear Creek Courant* newspaper in Idaho Springs starts at about 2:00 P.M. when the residents heard the town's fire siren let out a series of long wails. The owner of a local beauty shop ran to the newspaper and asked where the fire was. A reporter then headed to the firehouse and came across a patrolman talking on the phone. The man covered the receiver and, without being asked, relayed that someone was trapped inside "the Stanley." It was Cliff Morrison.

At the entrance to the tunnel, Sheriff Gene Kiefer was seen running out, flashlight in hand, and barking to miners and firefighters to put lights in the tunnel. He told someone that there was a man trapped under fifteen feet of dirt in the mine. Before hurrying back inside he said he didn't know if the man was still alive or not. A rescue attempt was fully underway.

An empty ore cart was used to shuttle dirt out of the tunnel as rescuers frantically raced to save the man's life. The reporter noted that a wooden door was discarded near the mine with the words "Keep Out" scrawled across it.

The sheriff later reported that they found the man but lost him again to a second cave-in. The men were desperately shoveling rock and dirt away from where Morrison was buried.

Sheriff posse captain and head of the county's mine rescue unit Al Mosch came stumbling out of the mine shirtless, exhausted, and gasping for breath. He put on a jacket to fight the evening's chill and collapsed to the ground. The paper described Mosch as someone who had mined in the county almost his entire life. Soon he was on his feet again and heading back into the tunnel. It's a day that Mosch vividly recalls. He remembers digging and trying to remove fifty tons of rock to save his longtime friend.

"We had a mucking machine in there and I got some of the volunteer fire people to help me because I couldn't get my team fast enough," Mosch said. "We got in there and took quite a few tons of rock out and his hand came up out of a pile of rock. He was dead."

Dealing with the shock and surprise of the moment Mosch did something he couldn't explain.

"He was a very close friend and I shook his dead hand."

The second cave-in happened soon after but not before Mosch marked the position of the body with a steel plate he had found. Morrison was finally removed after 5:00 P.M. The reporter noted the door was closed to the mine and a new "Keep Out" sign was put in place.

For Mosch this wasn't his first mine rescue, nor would it be his last. With more than a few close calls of his own, the third-generation hard rock miner has mining roots in Colorado nearly as old as the gold rush itself.

Once Upon a Time in the Rockies

Today in his mid-eighties, Mosch still digs in the mountains south of Idaho Springs looking for gold and occasionally silver. When he's not dreaming of the next big score, he operates one of his mines called the Phoenix as a tourist destination.

With a face that's witnessed a lifetime of difficult work, Mosch is still quick to share a grin displaying both his perfect teeth and the countless creases of time around his eyes. Indoors, or on the streets of Idaho Springs, it's almost like Mosch is perpetually stepping out of a dark tunnel into the light. A natural storyteller with all the essential charisma, Mosch is as close as they come in the former mining district to a celebrity.

His family has long chiseled out a living in the dark tunnels under the earth. His grandfather Rudolph "George" Von Mosch immigrated from Germany when he was sixteen years old, arriving in Colorado in the early 1880s.

Longtime gold miner Al Mosch stands in front of the Phoenix Mine. Mosch is a third-generation gold miner. *(Photo by D. Dahl)*

According to family lore, George walked from New York to Colorado on foot looking for an opportunity to make his way in the New World. With big ideas he first came to Grand Lake, Colorado, and homesteaded two hundred acres, hoping to start a resort similar to one his uncle had created years earlier in Wisconsin. As it turned out, fate had different plans for the Mosch family.

While cutting down trees on his homesteaded property he came across two groups of Native Americans fighting. "They started killing each other, the Arapaho and the Utes," Mosch said. Oblivious to him, George hid and watched as the violence unfolded in front of his eyes.

"So he figured he picked the wrong place for a resort. He abandoned it and came to Gilpin County and he learned all about gold mining and prospecting."

During World War I, George dropped the "Von" from the family name and temporarily changed his last name to Marsh to avoid the anti-German sentiment of the time. Mosch remembers listening to his grandfather's stories as a child and hearing how his grandfather single-jacked his way into the side of James Peak working a streak of gold ore.

George had a host of jobs over the years and eventually ended up as a lawman in Tolland, Colorado, where he also operated a pool hall.

"In fact, when I was a kid living up in Nederland, I think I was in grade school then, and he took me into a bar to buy me some soda pop," Mosch said. "I was drinking a soda pop and Grandpa was drinking a shot of whiskey and he had his six-shooter on him and his star and suspenders and another guy was drinking some whiskey. They got into discussing politics. Well, with more whiskey they got more hostile. It finally got really nasty and my grandfather took the star off his shirt and put the pistol on the counter and beat the hell out of this guy. He was eighty-two years old when he'd done that. That's the Western culture we don't hear much about these days."

In time Mosch's father, Hans, also became a career miner, working in fifty-two different Colorado mines. At that time hard rock miners used a compressed air-powered rock drill nicknamed by the men who used it "the widowmaker." Miners always dealt with the significant dangers of inhaling rock dust, but this mechanized drill made conditions much worse.

Miners often got silicosis from breathing in the silica found in the gold-rich quartz. Silicosis makes it all but impossible to breathe and led to a horrific death. Some operating the 150-pound drill succumbed to the disease in just six months.

There was no cure for miner's lung—except to change technologies. But that didn't happen overnight and silicosis became a widespread, deadly reality among hard rock miners looking for gold. One advertisement found in Central City from the late 1800s erroneously claimed it had a cure:

No More Miner's Complaint.
A New Medicine
The Miner's Drink

Mosch's uncle Walt was among those who fell victim to the silicosis-causing drill.

The Lamartine

The Dictator, the Tomboy, the Orient—there was no real rhyme or reason behind a mine's naming. Whether for luck, in honor of someone or something, or as an inside joke, the mines of Colorado hold a staggering variety of titles: there's the Afterthought Mine, the Dead Horse Mine, and the Home Sweet Home Mine.

The Lamartine Mine was named in honor of France's first romantic poet, Alphonse de Lamartine, who died in 1869. The mine was originally discovered in the 1860s but didn't go into production for another twenty years. Once up and running, it became a regular gold producer. Both Mosch's father and his grandfather Conrad Lahnert, a Russian immigrant on his mother's side, worked the Lamartine.

"It was a fantastic mine; it produced a huge amount of wealth. My dad worked in it, my uncles worked in it." And many years later Mosch worked in the mine's mill. As a child, Mosch's father told him stories of the places he'd worked. And one story stuck with him over the years. There was an incident in which some of the Lamartine miners "high-graded," or stole, a remarkable quantity of gold.

Mosch's father was popular among his fellow miners and was once asked if he wanted to join a small group who had come up with a clever way to remove an undocumented pocket of gold from the mine. Mosch said his father declined but swore to them he would keep both the story and the act a secret from the mine's owners. Keeping silent was likely a good idea as the plan's main conspirators nearly murdered one another on account of their accidental discovery of the rich gold vein.

"This is stuff that people never did find out about," Mosch said.

In the 1930s a Lamartine miner named Joe came across a cavity in the rock. About a mile into the mountain, he discovered a pencil-sized natural crevice running vertically and at right angles. A trickle of water flowed from it, suggesting an open cavity somewhere within.

Using his pneumatic drill and dynamite, he further widened the cavity and was stunned to discover what was inside. Joe revealed a large pocket of purest gold. The mine was profitable but the majority of the gold taken was in the form of gold ore, which had to be processed. A pocket of pure gold, including naturally formed gold wire, existed in a chamber of semitranslucent quartz crystals. The rich discovery was also totally off the map. Neither the mine's boss or owners were aware of its existence.

As hard as he tried, Joe couldn't figure out how to remove the large amount of gold from the mine without being discovered. In the end he decided to include his friend and fellow Lamartine miner Tony in the plot. Mosch's father told him the gold in the pocket looked like web spun throughout the chamber by "giant spiders." Tony then devised a plan to extricate the impossibly rich gold vein from the mine. And for it to work they would need additional manpower.

So Joe started yelling to a man named Herman, who was on a ladder repairing its bad rungs, which had given away recently and taken the life of one of the mine's workers. Herman was about one hundred feet up the five-hundred-foot ladder when he heard Joe trying to get his attention and urging him to come down. Herman became concerned that something serious had happened, so much so that he risked his life using the rotted ladder. He got to the bottom and followed Joe some three hundred feet down into the drift to where Tony was waiting for them in the dark. There he discovered the reason for alarm: enough gold to make them all very rich so long as the mine's owners didn't discover what they had.

The three figured they were relatively safe from discovery because the mining engineer Jones rarely made his way this deep into the Lamartine during his weekly inspections. Herman came up with a plan to place horizontal timbers along the side of the drift to hide the crack leading to the cavity of gold. The plan was then to assign loyal miners to sneak into the mine at night and work it after everyone had gone home for the day.

During this time a miner was placed as a watchman on the edge of the mine's dump overlooking the lower access road on which a moving vehicle could be heard approaching from a mile away. If the watchman raised the alarm, it should give the men enough time to hide their illicit activity.

High-grading required both an ethically flexible miner in addition to a dishonest assayer. It's said bartenders were often paid in gold and were never known to ask where the metal came from. It wasn't uncommon for miners to take just enough gold to pay their bar tab—but they never took large amounts. This was, in part, because the mineral was extremely difficult to remove unnoticed.

Miners skimming a little off the top could easily take home a chunk of rock equal to their day's wage. But because they needed to save up enough to make it worthwhile for an assayer, some slowly collected the valuable ore and hid it under their houses until they had enough. If a major vein was discovered it wasn't unusual for the mine owners, perhaps wisely, to hire guards and to pay hand-selected men more money to work the new discovery. But the miners of the Lamartine were sure they wouldn't be discovered due to the late hour of their work and depth of the location in the mine.

However, one day the mine's engineer, Jones, decided to leave his small, cramped office and go outside for a walk. He went to the mine dump and began wetting down some of the carload tons of waste rock recently taken from the mine to contain the dust.

The water washed away enough mud on one piece of rock to reveal a streak of gold running through the quartz. He washed another piece and took out his handkerchief to wipe it clean. It also held a streak of pure gold. Jones brought more water over to the waste pile and washed more pieces, discovering yet more gold.

The ore in the rock specimens he had found were worth more than the gold ore currently being taken from the mine. A lot more. Overlooking an odd piece of rock with a small amount might be common, but to dump out this much was downright strange. Jones found a hard hat and decided to go into the mine and find out where it had come from. He was going to get some answers.

Nearly one thousand feet below the surface he came across the three men working hard. They saw him too and couldn't figure who he was, with his headlamp swinging back and forth in time with his footsteps. The three had arranged things such that they were assured the chance to work in the area by themselves. The new tunnel they were working on was hidden behind mine timbers under the auspices of shoring up an area prone to cave-in. One of the three shined his light into the man's face and saw the mine boss. They couldn't imagine how he would have figured out their scheme, so they acted normally.

Jones complimented the men on the tight timbering job they had done in the area and asked them to come with him to try to find the source of the gold rich rocks being removed from the mine. Mosch's father told him the mine boss saw right away that their reaction to his request was off. Something was wrong. They asked him if he knew for sure that the gold rocks had indeed come from the mine. Jones said he did know that for a fact because they had come out of the mine's waste pile.

"Jones saw his men's discomfort; he knew something was amiss." He asked them how their repairs were coming along after a recent cave-in and whether they had to remove a lot of rock. He looked at the timbering and saw a rock piece similar to the one he had already come across. Jones decided to take a closer look. Joe, Tony, and Herman all knew what was coming next.

Among the men Jones was known as "disgustingly honest," and without talking the three prepared to kill him. Hans said each of the men picked up a large rock with which they could crush his skull. Perhaps they would later blame the engineer's death on an accident, something all too common. Jones was now on his knees and his back was to them as he examined the rock pile. They waited in silence to hear what he discovered and whether it would be his death sentence.

After a long moment Jones asked them to take a closer look at what he had come across. There was a lot of gold. Jones had a spark of an idea that caught fire as he was looking at the treasure. He told them he'd often asked to get them raises, but the mine's owners were greedy and would never allow such a thing. He said if they helped to keep the discovery quiet, he could use his expertise to remove the gold without anyone knowing and sell it. They could all be rich. He turned and saw the three still holding the rocks, although trying to hide them, and he was suddenly very glad he had presented them with the offer.

"That gold was never accounted for anywhere," Mosch said. "And the engineer had connections and he started what would become a major mining company with his profit from that one pocket." The story goes the men went their separate ways and became quite rich.

"The only one who didn't get rich was my dad. He knew all about it but kept his mouth shut."

A Hard Lesson

Growing up in a mining community, Mosch often skipped school to sneak down to where the men were working. "I'd go down at lunchtime if they were eating lunch and they'd give you a little candy if they had it."

Miners were often idolized by the local population because of their dangerous job and importance to the community. Even today the Clear Creek high school's sports mascot is a stout, mustachioed man in a miner's helmet who goes by the name of Golddigger Gus. Unquestionably, mining is tough and dangerous work. Mosch remembers the time his father was working the Gumtree Mine not far from the Lamartine Mine when tragedy struck.

In the early '40s, by himself and seven hundred feet into the Gumtree, Hans was working to get the remaining good ore from the mine. He spent his day mining and driving back and forth from the City of Arvada where the family had recently purchased a small farm.

His mother had a premonition that his father's life was in danger. She called the sheriff's office to go check on him at the mine. When he did, the sheriff found Hans unconscious just outside the main shaft. Hans was clearing an area in the mine about two hundred feet down a ladder when several tons of rock fell from farther up in the mine. The dangerous torrent of rocks took out the ladder and came rolling down into the area where Hans was working.

"Fortunately, the whole impact did not hit him—but he got plenty of damage from it," Mosch said. Badly hurt, his father then crawled back up the two hundred feet to the mouth of the mine. He turned to look back to see what had happened and then passed out. That's just how the sheriff had found him.

"When they found him up there he was close to death. He had sixteen broken bones, his skull fractured, his back broken, it tore up his guts," Mosch said. A week after the accident Hans almost didn't recognize himself. His hair had turned stark white.

"He spent almost a year rehabilitating in the hospital and got well enough and went back to the same mine by himself and repaired the damage. He got the thing working. He was tough."

A local newspaper article later reported that Hans worked the mine long enough to pay off his medical bills before finally closing it for good.

Sometimes living in a mining community could be rough and often very dangerous. Mosch recalled that when he was in grade school there was a local bully who tormented him. He said the boy once went as far as hitting Mosch in the back of the head with a hammer, knocking him out. He admitted that the boy terrified him.

"Back in those days we could take our blasting supplies home, before the new laws came out," Mosch said. One day the bully's father had taken home with him a bag of blasting caps. "The kid got a hold of them and he was swinging it around, and they blew up and blew his leg off and he died."

Mosch also remembered when he was a kid that his father had worked in a mine appropriately called the Terrible Mine. One evening he said his father came home for dinner in particularly low spirits.

"And usually when he came home at night Mom would have supper ready for him. He always had a good sense of humor and would relax and puff on a cigarette," Mosch said. "This one day he came back and something was bothering him. We were wondering what was wrong; my sister was little and I was little. Finally my mom asked him, after we got through eating, and he said, 'Yeah, my partner was killed today in the Terrible Mine.'"

His father explained that he had left his mining partner for a moment to refill his carbide lamp. When he returned, his partner was dead.

"Seventy tons of slab off a hanging wall let go and crushed him to death. I was told that there was no body to bury. There was nothing to take to a mortuary." Hans was required to get back to work the next day with a new partner.

One-Hundred-Pound Bucket

After a stint in the navy during the '50s and the few odd jobs that followed, Mosch went to his father and asked to be trained as a miner. "I was going to help my dad in the mines."

He said his father agreed despite his mother's disappointment. She never wanted him to go into that field of work, knowing all too well how difficult and dangerous it was. He spent long hours working with Hans, learning the tricks of the trade. The two first worked together in a tungsten mine called the Big Cameron, located in Nederland. Mosch said he lived in a trailer outside of the mine and worked twelve-hour days, seven days a week, for nearly a year.

One day while his father and mother left for Nebraska to spend time with his uncle, Mosch was left working the mine with just one other guy. Mosch would soon learn that mining didn't always draw the most sane individuals to its ranks. He said the man couldn't read or write and had the unsettling proclivity for eating recent roadkill—but all that mattered little as he worked hard and did a good job.

One day the two were working on some timber in the mine. Mosch was about ten feet above the other man when a hammer, weighing several pounds, slipped from his grasp and nearly hit the man below. Mosch apologized, but it didn't seem to matter—the man had taken it personally.

After a time they continued to work again and Mosch went one hundred feet down into a drift where he drilled seven six-foot deep holes, which were then loaded with dynamite. The plan was for Mosch to signal to the other man, light the fuses, and climb into a one-hundred-pound ore bucket. A 1927 Buick motor would then be used to hoist him and the bucket out of the hole and out of range of the explosion. Mosch made the signal and received the return signal. He lit the fuses, got into the bucket, and it started to lift him from the area. Then without warning, his ascent stopped about halfway to the top. He hung there for a time, unable to do anything as the fuses burned away. Then just as quickly, the bucket began to lower back to the bottom. In the hole there was nothing but smoke and the flicker of burning fuses. He didn't know what to do and was running out of time. Mosch said the only thing he could think of was to force himself as far as he could into the large ore bucket. The dynamite went off in sequential order. He said he remembers hearing the explosions, but after that everything went dark.

He woke to find that the man had hauled him up in the bucket after the blasts and dragged him outside, leaving him on the mine's waste dump. His hearing was seriously damaged, he was delirious, and he could barely move. He crawled five hundred feet back to the mine trailer. He said it took days to regain his strength and much longer for his hearing to return.

Later in town he came across the man again, who was surprised to see he was still alive. The man apologized and explained he thought Mosch had indeed tried to kill him with the hammer. However, desperate for the mine help, Mosch hired him back and said the two wound up working together again. Through all the years and in the many mines Mosch worked in, he said the majority of the men were loyal, honest, and not, at least overtly, homicidal.

Last Moments

Mosch was called on for his first major mine search-and-rescue operation in 1955. Two men had gone missing in the Marshall & Russell Tunnel near Empire. They were looking for uranium after discovering the right type of radioactivity in the mine's dump. They hoped to go in and discover the source. What they didn't know was that the mine had almost killed its miners many years before and was still quite dangerous.

In 1912 workers were drilling about six thousand feet from the surface portal when they came across what they thought was an underground river. They used several dynamite blasts and were clearing away the rock when there was a sudden rush of water. An underground stream filled the tunnel and violently swept the heavy ore carts in front of it, tumbling some over. The workers ran and escaped the mine. The news of the unusual underground river reached as far as Pittsburgh, Pennsylvania. The *Gazette Times* reported, "It is said this is the first occurrence of its kind ever recorded in Colorado." It certainly wouldn't be the last.

When the Marshall & Russell opened again forty-three years later, the water was found blocked by a deep natural dam inside the mine. Two men went in and found the air was bad, Mosch said, so they returned with breathing equipment, and using a raft to get across the water, they went farther into the dark tunnel. A third man waited for them outside and became concerned when they didn't return.

When Mosch and others rescuers arrived, they began to pump air into the cavern. Not far from the entrance the air compressor broke and many of the men decided to leave. Mosch and future Colorado mine commissioner Norm Blake remained and continued to work. They came across the dam and spotted the raft, but there was no sign of the men. Mosch said they broke the dam, releasing the water, and left to get oxygen masks. They soon returned and continued searching.

"By the time we got to them they were dead," Mosch said. They were found leaning against a rock, one with his mask off and the other with it still on. Mosch surmised the men's last moments as being trapped by the water and low on air.

"What happened is they went back into the mine until their breathing apparatuses were running low," Mosch explained. "Then they came back to get on their raft to come out and the raft was gone. They didn't anchor it."

The water was deep, cold enough to cause hypothermia, and vast, going on for several hundred feet. Mosch said their footprints indicated that they decided to try to find another way out. But they were unsuccessful. The commissioner of mines for the state of Colorado later recognized Mosch in a certificate for his "courageous and valorous efforts" during the rescue attempt. During another rescue sometime later, he had to wait in the mine for the coroner to come retrieve a body.

"I started to harden myself to that," Mosch said. "I was sitting on the dirt with his body at my feet, eating a sandwich, wondering why he didn't have more damage from being dead. I couldn't do that again—but it is amazing what you can do when you have to."

A Gold Mine

Mosch eventually got out of working in mines full-time and did it as a side job or on the weekends to supplement his work in the aeronautics industry. He and nineteen others guys did "exploration work" looking for gold in a mine near Idaho Springs up Oh My Gawd Road. Mosch said for many of the men it was a good excuse to get away from their wives for the weekend, do a little drinking, and blow off some steam. The mine had some good veins of gold in it, but after a time they called it quits. He met his future wife, a Colorado School of Mines student, at that mine looking for a tour.

That mine, like many hundreds of mines in Clear Creek County, has since disappeared, all but eroding away into obscurity. Eventually he decided to get back into the industry and worked in Utah, mining uranium.

"We needed uranium to make atom bombs to protect us from the Soviets," Mosch said. "They were doing the same thing and protecting themselves from us." Mosch returned to Colorado and Clear Creek County and worked in the Senator Mine in Dumont, the Poor Man Mine, and the Brazil.

In 1972 he bought the Phoenix Gold Mine and worked it for years. In the late '80s after giving a tour to a group of veterans he decided to open the mine as a tourist attraction, while still doing exploration work on the other mining properties that he owned.

Mosch said he doesn't know why gold mining brought him back year after year, especially when it meant risking his life.

"It's just an obsession," he said. "I'm stuck with a mining fever." Even now in his mid-eighties, he's planning to do exploration work on a forgotten gold vein connected to the old Lamartine.

"I've held [the claim] for a long time, always had it in the back of my mind." And maybe, like the main characters from his father's tale, he'll walk away a rich man.

CHAPTER 7

THE NEW PROSPECTORS

As a kid, James Long remembers traveling with his father across unforgiving southwestern Colorado during the 1950s. After serving in World War II, James's father felt lost and looked for something to keep him busy. Accustomed to hard work, having grown up on a ranch, his dad eventually gravitated toward prospecting. It being the Cold War, his father took him to look for what the government prized at the moment even more highly than gold—uranium.

"Back in those days I was more interested in wildlife and flowers, and I wasn't interested in gold or silver, and he wasn't looking for gold or silver. He was looking for uranium, and I knew that stuff was dangerous because he said it was," Long explained. Geiger counter in hand, his father looked for uranium deposits on the claims he owned.

"I didn't pay a lot of attention to the stuff he did, but he had a number of claims in that area, and he found what he was looking for."

The experience left an impression on Long and years later, the prospecting seed was still planted in his mind. After a long career in law enforcement he, like his father, was looking for something to do. Long had all of his dad's equipment, books, and notebooks but wasn't interested in searching for radioactive rock in the southern portions of the state. He had grown interested in gold, the hunt for the metal and its lore. Over the years, he had read about gold prospectors in books and found that searching for it appealed to him.

But there were no classes or teachers, and Long learned to look for gold the hard way. By the time he finally retired from law enforcement, he knew the tricks of the prospecting trade and was ready not to only invest his time in searching for gold but to teach others how to find it.

Dying for Gold

It was early evening, and tucked away from the majority of the noise and neon of Denver's busy Colfax Avenue was a small community center. Inside, the building's lights were dimmed and its visitors were pointed either to the left or the right with two handmade signs: one advertising a tai chi class, and the other a meeting area for the Gold Prospectors of the Rockies. Down a hallway, more than eighty people filled a large meeting space. They walked around and chatted with each other waiting for the meeting to start.

One man leaned over to an acquaintance and said over the din, "I just got a few flakes, how about you?" The other shook his head discontentedly. Despite the large crowd, it was a slow night for the club, which brings as many as 150 members to its monthly gatherings. The age range of those present varied, but the majority were male, with graying hair and wearing baseball caps. There were cookies on one table, and recent treasure-hunting discoveries on another. People gathered around the "treasure table" to vote on which recently discovered items they thought were the best.

The meeting soon started and people filed neatly into the rows of metal folding chairs, taking their seats as Long appeared, ready to address his audience. President of the organization since 2011, Long wore a neatly trimmed silver goatee and a faded camo-colored NRA baseball cap. He stood behind a wooden podium; directly behind him on the wall hung a giant American flag. After a brief introduction, the meeting kicked off with a documentary highlighting the conflicts throughout history caused by gold. The audience watched quietly; the only sound was an occasional cough or the hiss of an oxygen tank used for breathing by some of the more senior members. After the movie ended, Long returned to the podium.

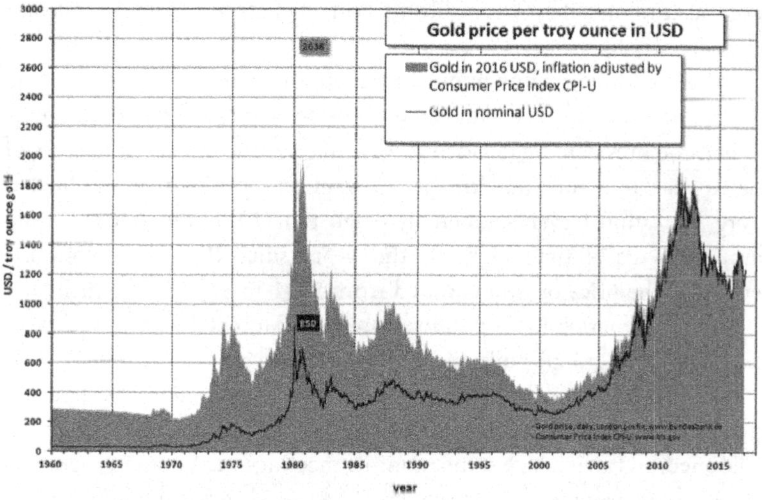

The price of gold constantly fluctuates, but it reached its all-time high in August 2011 at $1,895 an ounce. *(Courtesy of www.bundesbank.de and www.bls.gov)*

"A lot of people have died for gold; don't you be one of them," he said. Then after a pause for emphasis, he went into a description of the day's turbulent precious metal values.

Following in the muddy footprints of the prospectors that have come before, these new gold-hunting enthusiasts are trying to supplement either their wages or retirements with the yellow mineral. Some do it for fun, some for profit or addiction. And there are those who look for gold just because it's there. The gathering felt slightly evangelical in nature, which is appropriate because for many of those who attend these meetings hunting for gold is a religion.

No Hobby

When Long joined the organization almost ten years ago, the membership of gold enthusiasts was small. Those who were members then and are still today he calls "old-timers."

"These were guys who had knocked around, they had dredged, they had prospected a lot of Colorado, and they found a lot of gold," Long said. "They had the opportunity to work back in the days when there was a lot more access, and they were able to find nuggets and some significant amounts of gold."

He said the group's public outreach was limited to a few panning demonstrations a year. He gravitated to the organization's board, and once he became president he decided it was time for the Gold Prospectors of the Rockies to branch out. The group went on more outings and hosted more educational events. Then in August of 2011, the price of gold jumped to its all-time high. In the years since the 2007 economic recession, the price of gold climbed from $841 to $1,895 an ounce. The attempt at garnering more public awareness combined with the price of gold caused the group's ranks to swell.

"People were looking around, but they weren't finding an organization that was doing anything," Long explained. Membership ballooned, and attendance at the monthly meetings went from 35 members to 135. "Our membership dramatically increased overnight, and we had dozens of people calling us," Long said. One couple called Long from New Jersey saying they had quit their jobs and planned to move west to Colorado to look for gold. They were hoping to strike it rich and asked him where they could find gold.

"I said, 'Listen closely, because I'm going to tell you the truth. If I knew where you could go to get enough gold to retire, I wouldn't tell you. Because I'd already be there and I'd have all the gold for myself. And that's the nature of prospecting. You need to understand that.'" He then told them that he recommended they get their jobs back.

"'You'll make a heck of a lot more money, and you'll retire a hell of a lot more comfortable than you will if you come to Colorado and try to prospect for gold.'"

And although the price of gold dipped down in the following years, the group's membership continued to grow.

"We're dedicated to the amateur prospector. We're more in the teaching mode and people from the outside that want to get interested in small-scale mining can come to us and learn," Long said. The term "small-scale mining" is of particular importance to him.

"We don't like that word 'hobby' because hobby causes all kinds of problems," Long said. He added a hobby doesn't lend itself well to digging for gold. Instead, it opens up would-be prospectors to increased scrutiny.

"We see it not as a 'hobby' but as recreational small-scale mining," Long said. "If you want a hobby, build model airplanes. Everybody is happy with that but when you go out and start digging in the ground people kind of frown and go, 'What are you doing?' They don't understand, so you have to explain to them what you're doing and demonstrate and show them that you're not destroying the habitat, you're not destroying the ecology, you're not destroying the environment, you're not polluting it."

The group works hard to counter any negative associations with what they're doing by having trash cleanup days, removing from the streams mercury left over from historic mining, and removing broken glass.

Not All That Glitters

Before he retired, for his first serious foray into gold hunting, Long traveled to a stream and panned out and put into jars what he was thought was gold.

"I was up in the Arkansas River drainage and having a good time, and I found gold," Long said. "I spent four hours, and I had two pint jars full of gold. And man, I was all kinds of excited." He was thinking he'd have to go someplace in Denver to find out how much it was all worth when an older man with white hair came down the bank and asked him what he was doing.

"I said, 'Well, I'm digging for gold' and he said, 'Well, you having any luck?' And I said, 'As a matter of fact I'm doing quite well.'" The older man asked to look at Long's findings and then had the unfortunate duty of breaking the news to Long that he hadn't found gold at all.

"He reached out and put his arm on my shoulder, and he said, 'Son, I've got bad news for you. What you have [there] are pyrite and mica. We call it fool's gold.'" The man then offered to show Long how to find gold, and explained how it was challenging work and what little gold was found tended to be in tiny quantities.

"And one hour with the old gentleman changed everything for me," Long said. "When I found that first gold in my pan it was five little specks—I was floored, it was beautiful. They were not huge, but they were plainly yellow and plainly gold."

Since that day Long has dredged, high-banked, panned, and sluiced his way across the state to find gold. He's gone in the summer and the winter. Once he took an ax to a frozen stream to break up the ice so he could look for gold beneath it.

"I've had a great time, I've found quite a bit of gold, I have a fair amount of gold at home, but I have never sold a single gram," Long said.

And he's not alone. He said he knows one prospector who owns at least fifty ounces of gold, which at its current price of $1,224 an ounce in 2016 was worth roughly $61,000. Long said the man keeps all his gold in a safe, never selling any of it.

"For him it is not about the selling—it is about the finding and the having." He added many people don't understand where gold comes from, what makes it so unique, and why even the smallest amounts of the metal are special.

"Once they understand what processes occurred in the very beginning for gold to even exist on our planet and then get to the point where you can find it, it's amazing," Long said. "They have no idea where the gold came from, how it got to be here, how it got to be in the streams. They don't understand those processes."

The Processes

Bearded and wearing a Grateful Dead T-shirt and a faded blue jean jacket, there is likely no better person to ask about the rare yellow metal than world-renowned expert on the subject Dr. Richard Goldfarb. Some weeks before the Gold Prospectors of the Rockies met, and just days before he was scheduled to fly to China to teach on the subject, Goldfarb sat with me in a little coffee shop in the City of Golden to discuss the nature of the elusive mineral.

"There's a lot of theories about how it got into Earth," Goldfarb said, tackling the subject with enthusiasm. "When the Earth formed four billion years ago it [gold] may have been in the core of the Earth. Alternatively, other people have argued that it came with a meteorite collision four billion years ago."

Goldfarb didn't originally intend to be one of the world's foremost experts on the metal. He received his undergraduate degree in geology

World-renowned gold expert Dr. Richard Goldfarb. *(Photo by Chancey Bush)*

from Bucknell University in Pennsylvania before studying hydrology at the University of California, Berkeley. He later dropped out of school and "lived on the streets for a few years" before he earned his master's degree in hydrology at the Mackay School of Mines in Nevada. He was later hired to run the US Geological Survey's (USGS) mineral exploration program in Alaska. At the time, he didn't know that much about minerals but thought Alaska "sounded like fun."

"So I got in my Pinto and drove here from Reno. They hired me; I got a couple of students who came with the project. We shipped all of our alcohol up to Alaska because it was back in the old days when Anchorage had dirt streets, and it was filled with bar fights and prostitutes and everything—it was great!"

Later in his spare time he got his PhD in geology with a focus in gold. It was then he changed the model for looking at how gold forms along the Pacific margin of North America and Alaska.

"I tried to avoid management as much as I could because I love doing science. So I was in charge of a lot of our work in Alaska's mineral program."

After thirty years of working for the USGS and heading up a

department, he retired and now works as an adjunct professor at several universities around the world. Goldfarb said however gold came to be on Earth, what happened next is not in question. As the planet began developing its core, gold began to accumulate in different ways in the mantle and the crust.

"We just couldn't go out and mine just any rock on the surface for gold," Goldfarb said. "Certain processes in the Earth's surface and in the shallow crust concentrate gold in veins or in disseminations, and that is how we mine it."

As mountains began to form, in places like Colorado, California, and Nevada, so too did orogenic gold deposits. About thirty million years ago as granite began to develop, it did so with gold-rich fluids forming in veins. Underground water with gold and quartz particles in it was forced into the rock by tremendous pressure. Gold then crystallized and was encased in quartz. Goldfarb said this happened in the upper two kilometers of the Earth's crust. These veins of gold are sometimes exposed to the elements and natural erosion that washes the gold from the mountains into the streams. These alluvial gold deposits collect in streams, becoming what the miners call placer deposits.

"When the Rockies uplifted, some of the granite-related gold went downstream and concentrated. So it's the uplift of the mountains and then the eroding," Goldfarb explained. "You need to expose it at the Earth's surface so that it can erode away. Once the veins reach the Earth's surface, they start to weather and then erode." Goldfarb added the gold deposits connected with granite were located in relatively young rock that hasn't yet completely eroded away.

"Most of Colorado's gold is thirty million years old," Goldfarb said. When early prospectors found gold in the streams on Colorado's Front Range, they followed it up into the mountains to the source. Goldfarb said there is currently a lot of debate as to how that alluvial, or placer, gold forms. For instance, the famous gold nugget isn't formed in that state in the veins. The thought is that the gold flakes are squeezed together in the streams to form the nuggets.

"But there's a lot of evidence that sometimes . . . [gold] comes out a solution around bacteria or some sort of an organic material," Goldfarb said. Essentially the premise is that gold precipitates from a solution onto bacteria/microorganisms rather than on just the preexisting gold grains in the streams.

"The theory is that certain bacteria species can simply accumulate gold from [a] solution under the correct chemical conditions," Goldfarb said. He added this theory of how gold is formed is still hugely controversial, but there is plenty of research suggesting that this is a very real possibility.

No Fortune

Back at the community center Long quizzed the audience about where gold was first found in Colorado. "Montana City," many answered back. He then asked where it was found next. "Auraria," they responded, and then "Arapahoe City." Long continued to ask questions and the audience answered, as familiar with the locations and history he references as the backs of their hands. The group knows gold very well.

Long told them about an upcoming visit to Montana City, now located on the east bank of the South Platte River and a small park called Grant-Frontier Park. He said that a group of construction workers agreed to allow the club access to look for gold there. He said he scouted ahead and found a little gold but added finding it in that location was more historically significant than monetarily. Long then talked about discovering other things while out looking for gold, such as antique watches. Someone yells "cell phones," and he agreed, adding he found a pair of dentures once. Then they began the raffle.

For the group's members, even attending the monthly meeting is all about the chance of getting real gold. There are contests, gold-panning events, a hidden cache where clues are presented each month, drawings, and raffles. But the raffle for gold nuggets is the evening's highlight. As it began, there's a strained silence and some members clutched their lottery tickets in anticipation. A man standing up front, clearly understanding the importance of his position, gravely pulled a card from what looked like a battered, and presumably clean, spittoon. During the first raffle, a woman was selected. Some members joked that she always has good luck during the raffles. Then many up front groaned as they learned she got the biggest nugget of the night. She walked back to her seat with a spring in her step.

"Whatcha get?" one person, possibly her husband, asked. "I got a nugget," she answered with a smile.

For many, this night can provide them with more gold than they'll easily find in nature. Long believes many of the state's streams were picked clean more than a hundred years ago. But gold is gold, and whether it is found under a boulder or won in a Wednesday-night raffle, the thrill of the hunt is just as real. After one man's number came up, he shouted, "Oh yeah, first one in two years!" Wearing a turquoise-colored shirt and a baseball hat he went up to collect his winnings. Later his number was called a second time, and after another shout of joy it was clear he could barely contain his exuberance. Long dismisses the idea that the average person can go out and find a fortune's worth of gold in a stream in this day and age.

"For the most part, what you're going to do is go out, work your butt off, and you're going to spend a $100 to find $10 worth of gold," Long said. "There's always that rare chance you're going to discover or tap into an undiscovered vein or find a pocket of gold. There's always that chance but when you compare it to percentages, the percentages are extremely low that that's going to occur."

He knows people who have prospected for gold for twenty years and never found more than an ounce over the entire time. "They're never going to become rich at mining like they did in the old days when they could come out here and . . . the streams were full of nuggets."

It does then beg the question that if someone can't find enough gold to become rich, why do it at all? Why spend the hours hunched over a stream in all weather conditions looking for something they're not likely to find or indeed find very little of?

"I do believe it is the lure of gold," Long said. "When you find that first gold in your pan and you see that beautiful yellow, and you realize that you are the first person ever to see that gold and it is yours. You found it, and it is something that has been held precious in the eyes of man literally since the beginning of time when it was first discovered. The fact is gold is beautiful. Gold is something you have to work for; you just can't find it lying around anymore. When you find it, it's special and if you find enough of it—it has an intrinsic value and people recognize that."

With the sound of scraping chair legs, the Gold Prospectors of the

Rockies meeting came to an end, and its members prepared to head home for the evening. They filed out into the community center's hallways and opened its glass doors that led into the parking lot beyond. Some of them thought about gold and a lucky few left with a nugget or two of the precious metal.

"A lot of our members, now I'm finding, are the retired types, and so this is a way recreationally they can become a small-scale miner in their own right and say before they die they prospected and found gold. They became part of the history of this state."

Just north of the community center lies Clear Creek, and thirty miles west along the stream is a mining claim owned by a man not interested in the past, only the future. And after a lifetime of prospecting for gold, one day, Chad Watkins plans to strike it rich.

CHAPTER 8

A GOLD TOOTH AND A PAIR OF PISTOLS

Underwater and looking for gold is the perfect environment for Chad Watkins. Using a suction hose, he pulled up gravel and dirt that he sent to the surface where his dredge floated and worked to filter out the gold. Deep below the river it's a world with virtually no noise. The "tut tut tut" of the dredge's motor doesn't reach him here.

"There's nothing I love more than being underwater dredging—it is meditation," Watkins said. "It's you and the river."

On a clear day, he can see a hundred yards upstream. He can see the fish on the bottom, but the traffic, the people—all that is forgotten. Working a hole, he reached for bedrock with his hose often working around and beneath giant rocks. While calming, it certainly isn't safe underneath the surface of Clear Creek or other Colorado streams, where Watkins spends his time hunting for the elusive mineral.

"You'll hear the big boulders when they drop out of the wall," Watkins said. Those boulders can pin a gold prospector to the bottom of the streambed, crushing the life from him.

"You've got a hand on it, and I've got my feet floating, so if the rock goes, it pushes me back and doesn't pin me," Watkins said. "My hands look like I've been in a fight just from rocks and nozzles and just banging and bruising."

But all the danger doesn't necessarily go away when he decides to go back to the surface. Sometimes he emerges from the cold water to discover people waiting for him. Waiting and watching to see what he's found and looking at the expensive prospecting equipment lying on his gold mining claim.

"I used to keep a shotgun on the inside of my dredge, where you couldn't see it, especially when I was working by myself," Watkins said. "I came out of the water one day, and there are three guys standing on the bank looking at me, looking at my gear. I shut off the machine, took off my mask, grabbed the shotgun under from where I had it clipped, and asked, 'How's it going?'"

The men backed up without saying anything and left. Looking for gold is dangerous, and it's not always just because of the inherent danger of mining, or working with boulders underwater. But it's a job Watkins has no intention of ever giving up.

"They'll find me dead on the side of the stream one day," he jokes.

Watkins has spent the better part of twenty-six years looking for gold and is something of a celebrity gold prospector. An outfitter, teacher, and guide—he was recently featured on several seasons of the reality television series *Ice Cold Gold* on Animal Planet. The show follows prospectors as they hunt for precious minerals across Greenland. When he's not looking for gold, he's working odd jobs to fund his gold mining activities. And he finds gold. Watkins believes his passion for it is in his family's heritage. He finds it waiting around the next bend and behind the next boulder.

"It is the lure of that next pan, or turning over that next rock and a fortune is going to be laying there for me—I'm a fourth-generation native, and Colorado history is in my blood."

Tenacity and Laser Tag

More than twenty years ago Watkins was working in a family entertainment business that specialized in laser tag. While not exactly extinct, the game and its technology are synonymous with the 1990s and pizza parties. The sport often saw teams of people running through neon-splattered mazes and shooting at one another with futuristic-looking guns.

Outfitter, teacher, and guide Chad Watkins stands next to his dredge while looking for gold in Greenland. *(Courtesy of Chad Watkins)*

"Whether I was operating a store, or working on the manufacturing side, we'd get home late at night, usually after drinking," Watkins said. "Back then you'd come home at 2:00 A.M. when the only thing on TV was the commercials for the Gold Prospectors Association of America."

Watkins said the commercials offered membership, a gold pan, a snuffer bottle, and access to local gold mining claims. Night after night he came home, turned on the television, and saw the same infomercial.

"And apparently one night I ordered it—I don't remember it," Watkins said. "One day it showed up in the mail, and I'm like 'all right.'"

As a kid, Watkins had explored some of the mines near where he grew up in rural Colorado, something he admits to as not being the cleverest thing he's ever done. Although he saw that mining and its history were all around him, he hadn't actually considered giving it a try himself until it arrived at his front door. But whether it was providence or just luck he took the gold panning supplies and headed into the Rockies to find gold.

"It took about two weeks before I really saw any color," Watkins said. "But once I found a little bit of gold and learned that it was out there—that was all it took."

While working Clear Creek, where he would later share a claim, he met an old-timer named Charlie who was "an old tramp miner," or someone who traveled from place to place plying their underground trade.

"And he was just a mean old bastard," Watkins recalled. "But if he liked you there's nothing he wouldn't do. We got along really well. He took hundreds of ounces out of that claim. I saw him."

One day Charlie, then in his seventies, invited Watkins across the river to sit and talk.

"We'd just sit out there and talk and tell stories about gold and how to find it," Watkins said. Over the next nine years, he spent his summers, often with Charlie, prospecting and digging for gold. During that time, he remembers one day finding a solid ounce worth of gold in four hours.

"Those are the days when you get ounces in your box . . . and then days when it disappears," Watkins said.

He said the key to looking for gold was to find someone who knew more than you did and learn as much from them as possible. As such, Watkins took as many notes as he could from the old-timer, adding Charlie's best advice for looking for gold was essentially to be stubborn and never give up.

"You don't stop, you keep looking—go another two feet, and you may hit it, and if it is not there you go another foot, and you might hit it," Watkins said. "It's just tenacity. It is there, but you got to look for it. There's nothing easy about it—and it doesn't matter if you're working hard rock or surface mine or dredging: it is all dangerous."

Dangerous

Watkins said that sitting by the side of the stream with a gold pan is a pretty safe activity for most people, barring any unforeseen close encounters with giant or particularly aggressive wildlife.

"But I've seen people when they're digging in the riverbanks, they're tunneling in five or six feet, and not knocking that overburden down, and those things collapse on people," Watkins said. "I know of people who have been killed over the years and people who have been close."

Watkins said some prospectors get buried alive and are lucky to have a gold-prospecting partner nearby to grab them by their kicking feet to yank them free. A good mining friend of his died in 2012 at the Ajax Mine, located near Colorado's Mosquito Pass.

"He was shoring up a section in winter, getting ready for production, and it is a rich mine," Watkins said. "A boulder came down out of the dike above, pinned him in there, and he was in there about three days, they figure, before he died. His brother went up and found him; he hadn't checked in for a while."

Of course, mining for gold in rivers is also dangerous.

"During high flow you stand out there, and you can hear those boulders banging down the creek bed," Watkins said. Or standing in a hole and watching as you're suddenly in the middle of a 360-degree underwater landslide.

"Sometimes you're digging down, and you've got a head wall, and you've got a four-hundred-pound boulder above your head that you've got to control as it falls into the hole," Watkins said. Given the obvious danger of looking for gold and what is, for the most part, a series of diminishing returns, I had to ask why he'd risk serious injury or even his life for an ounce of gold?

"Stock car racing is dangerous," Watkins said, shrugging. "It is just one of those things you get a passion for." As a gold outfitter, Watkins said one of his jobs is to manage the expectations of his clients.

"The first thing is I tell people 'We're going to come out, and we're going to find some gold.' I never promise it, but I'm fairly certain I can put you onto a spot and with a little bit of hard work you can find some color. No doubt about it," Watkins said.

But the likelihood that they're going to find thousands of dollars is slim. Watkins then added without hesitation that people can still get rich looking for gold in the streams of Colorado. For his part, he said he's done quite well some years and others not well at all.

"That's just the nature of it. You get into a hot spot, and you're into thousands of dollars sometimes in a matter of hours," Watkins said. "There are days where you're digging and digging, and you'll hit a pocket or a nice little pay streak, and you'll pull two or three ounces out, and you go back the next day, and there is nothing. It is gone. Gold, it makes you work for it."

He said that for himself he can move in and out of gold prospecting from a hobby to a living and back again.

"I've got friends that do a little at a time, and they are happy with that pace. They're outside and find a little gold and hanging out with their friends," Watkins said. "For me when it is time to turn the machines on: 'We're not talking now, shut up and dig.'"

When he was recruited for the reality show *Ice Cold Gold,* he said there was a lot of walking and it often reminded him of a scene from *The Lord of the Rings* with a bunch of guys hiking across the Greenland landscape. He was eventually kicked off in typical reality TV show fashion. But just because he's no longer being filmed doesn't mean he's abandoned the quest.

"Most of the easy gold has been found. There are places where people haven't looked. The floods bring in the new gold and bring stuff out of the gulches," Watkins said. "Once the summer gets here you spend as much time as you can trying to get at it." For Watkins looking for gold isn't just about finding wealth but also getting in touch with his and the state's history.

"It's sitting out on a spot knowing someone was sitting there 150 years ago . . . the fortunes are out there. It is just getting to that spot to find it," Watkins said. And some days both history and the gold become one. Not long ago Watkins found what at first must have looked like a gold nugget. After further inspection, he realized it was a gold tooth, with a bit of bone still attached to it. His mining partner discovered a pair of old revolvers just a few feet upstream.

"There's a story there," Watkins said.

While no one will ever know for sure, Watkins suspects a prospector during the gold rush might have been shot by someone looking to appropriate his equipment and ended up dead in the stream. What isn't a mystery is that the discovery of gold and its unimaginable wealth, perceived or otherwise, brought with it crime, violence, and murder.

CHAPTER 9

GUNSLINGERS, KILLERS, AND GHOSTS

Gold was located on the far western edge of the Kansas Territory when the rush for wealth ignited the frenzied dash across the Great Plains. Colorado wasn't a state, or even its own territory, when gold was discovered. The stories from this era are often chaotic and violent, creating a window into a time when gun smoke drifted above the towns fertilized by the sudden discovery.

It was evident the law of the country didn't exist in any official capacity, and people didn't play nice for long. Justice was dispensed through vigilantes, bounty hunters, and, often as not, found dangling from the end of a hangman's rope.

Two Hangings and a Shootout

"You're ghouls, all of you are ghouls," Christine Bradley jokingly accused the standing-room-only crowd.

A predominantly silver-haired audience shuffled in from the cold early evening of Georgetown to sit in row after row of metal folding chairs in the community center. They were there for Bradley's 2016 early spring presentation on behalf of the Devil's Gate History Club.

The county archivist said she was recently helping to clean out a small storage room in the county courthouse when she came across ten boxes of coroner inquests starting in 1867.

"I'd never seen these records before and started pawing through them going, 'Oh, this is interesting stuff.'"

What she found consisted of 278 cases of deaths in Clear Creek County requiring a coroner's report, investigation, and subsequent interviews. The discovery gives a genuine look back at a time when a functioning legal system was more of a lofty notion than a stone-cold reality.

In those forgotten boxes she found seventy-two mining-related deaths, by far the biggest category; thirty-one suicides; and eighteen murders. Out of those homicides, thirteen were caused by gunshot, four from stabbing, and one man was beaten to death in a bar. She added a great number of deaths could also be described as suspicious or accidental, but ultimately didn't get much in the way of an investigation.

"But you have to remember that coroners' inquests in the nineteenth century were done in a time when there was no refrigeration. The only refrigeration up here was in the months from November to February," Bradley said. "So things had to happen pretty quickly—or it was kind of gross."

Bradley said when a questionable death occurred, a coroner would gather up a jury to go to the site of the death and interview the witnesses and those involved.

"I wonder what it would have been like to be the friend of a coroner back then? 'Hey, I got another one.'"

Even when law and order had found its first tentative footholds in the early Colorado mining towns and camps, frontier justice was faster and often realized in the form of an angry mob dragging a rope.

On April 24, 1867, Ed Bainbridge was playing a game of cards in the Nickel Saloon. By this time Georgetown had already turned its original focus from gold to becoming one of the West's wealthiest silver towns. The story goes that Bainbridge was playing a man named Jim Martin and soon the game started going south.

"I think they were playing for some serious money, and our friend Mr. Bainbridge was not doing well," Bradley said. Bainbridge continued to get the worst of the game and at some point got up from the table, removed his pistol, and shot Martin in the face.

A stereoscopic photo of Georgetown from 1894. *(Courtesy of Christine Bradley)*

He was then restrained and confined to the top floor of a house so that he could later be tried. Martin was expected to die from the rather nasty wound inflicted by the bullet. An angry mob formed, and like so many red ants, swarmed the house Bainbridge was held in, accessed the second floor, tied a rope around his neck, and dragged him screaming through the window.

Bradley said a local paper reported, "An excited mob broke for Bainbridge last evening and ran him up a tree from whence he dropped down a corpse."

He was found the next day, still hanging from a tree. It turned out, however, that Bainbridge's victim, Jim Martin, did not die.

"The fact that he was hung for murdering a man that didn't die was not a good thing," Bradley said, questioning whether Bainbridge's multiracial background factored into the mob's bloodlust. "I think that's why it's a long time before there's another hanging in Georgetown."

Only months later some of the town's residents swore that Bainbridge's ghost came back to haunt them.

"The Ghost of Bainbridge, who was hung by a vigilance committee, has returned to earth to vex and worry the people who lived in a house . . . by the fatal tree," stated a reporter for the *Rocky Mountain News.* "It opens doors that are locked and slams them in a way supposed to be natural for a house breaker. One citizen of keen perception has seen him with the rope still around his neck. The family has moved away."

Georgetown's second vigilante execution occurred on December 15, 1877, when Robert Schramley confessed to murdering the town's butcher by Georgetown Lake. He then tried to escape but was caught, and justice was, once again, handed down sans a judge or jury. When he was found, there was a note pinned to his chest.

> *Vigilantes around!! No more murders!! Behold the fate of this man, the same terrible end awaits all murderers. Life and the public security is too sacred not to keep protected, even by resort of the unpleasant means of lynch law. Take warning, take warning. Else ye murderers the fate this brute Schramley has met with awaits you.*
>
> *By the order of committee of vigilantes.*

Gunfights are popularized in movies and books of the Old West to the point where now one can hardly imagine what it was like to get groceries—without having to shoot down at least two or three black-clad desperadoes. And while gunfights obviously weren't quite as common as is often believed, and likely most were more myth than reality, they did on occasion occur.

In Mill City, near present-day Dumont, Colorado, between the mining towns of Georgetown and Idaho Springs, William Dowd was looking to get drunk and kill himself an abolitionist. As far as plans go, it might have been a bit shortsighted, as the year was 1867 and the War Between the States had ended two years before.

A witness reported to the coroner's jury that Dowd was in the local saloon and regularly calling for drinks and talking politics with the patrons. Soon his mood soured and he became intoxicated, threatening to shoot any abolitionist he came across. If there were indeed any in the room, they didn't immediately speak up. Witnesses said Dowd had a pistol in one hand as he made his threats. A man called Isaacs entered the bar and the two drank together and discussed the political issues of the day. The witness said they drank an additional fifteen to twenty drinks apiece, a room-spinning, stupor-inducing quantity to be sure.

Perhaps unsurprisingly, Dowd soon became agitated once again, drew his pistol, and once more threatened to shoot any abolitionists in attendance at the saloon. A brave and quick-thinking onlooker took Dowd's gun away, put it back in his holster, and suggested Dowd continue his journey on to Denver. Dowd declined and more drinks were called to help pacify him. This apparently didn't last, and Isaacs drew his own pistol over a dispute with Dowd, encouraging him to join him in a duel.

A bystander said, "[Dowd] drew his pistol and stepped to the window, I was afraid . . . and stood behind the desk out of sight. Three or four pistol shots were fired." Having been shot, Dowd then walked out of the saloon and fell to the ground dead.

While murders seem to have been familiar enough, Bradley said the shootout between the two men was interesting because rarely did gunfights break out in the mining towns. Gold towns did prove to be a little more violent than those that transitioned to silver. Miners carrying the gold dust with them and using it to pay for supplies sometimes tempted the unscrupulous, and that brought a world of violence.

CHAPTER 10

THE TALE OF TWO CITIES

A prospecting group from Kansas established Montana City, Colorado, at Dry Creek in the empty diggings left by William Green Russell and his party in 1858. It must have been disheartening when they found what little gold that was once there was now long gone. The city died without a second thought when its residents agreed as one to move farther downstream to a better and potentially more fruitful location. In this new spot they created the St. Charles Town Association. As the winter of 1858 now blew across the plains and collected along the sides of their crude cabins, the majority opted to go back to Kansas for the winter, leaving only a few men behind to brave the cold.

In 1859 when the Russell party returned, they found the small town of St. Charles waiting for them. They decided to create their own town called Auraria, after a town in Georgia that was the site of an earlier gold rush. Auraria is derived from the Latin word for gold, *aurum*. Wishful thinking, as any of the real gold was still hidden deep in the mountains. But money could be made from the town itself: shares could be sold, and rent and sales earned from properties.

A third group came in and created Denver City, named for the governor of the Kansas Territory, over the site of St. Charles. The Denverites essentially strong-armed the former townsite to build their own. Legality in the far reaches of the territories, such as property rights, came only as a hazy, distant afterthought. This is especially apparent when recalling that the land on which the towns were built legally belonged to the Arapaho.

Today all that remains of Auraria is a college campus and some quaint Victorian-era houses on the southern side of the school, bordering Colfax Avenue, a road once synonymous for its ill-begotten and well-deserved reputation. Denver currently dwarfs the former town and surrounds it like a fist. But during the late 1800s, Auraria and Denver were equal and bitter rivals. One of the area's first killings further inflamed the tensions between the two cities.

Peleg T. Bassett, Denver's first town recorder, and John Scudder, the treasurer of Auraria Town Company, were once good friends but soon became implacable enemies. On April 16, 1859, Scudder heard Bassett telling others he wasn't trustworthy and accusing the Denver recorder of opening his mail. An insult too serious to ignore, Scudder went to Bassett's cabin to confront him about what he claimed were false accusations. Apparently Bassett initially denied bad-mouthing his former friend, but as their conversation grew heated, he admitted that he had indeed spoken the unkind words.

The argument grew more intense and Bassett grabbed a pick handle and lunged at Scudder, who evaded the attack, drew his pistol, and shot Bassett at close range. Following the discharge of the firearm, Bassett stumbled back against his house with a bullet in his chest and soon died. Scudder tried to say his actions were in self-defense, but a lynching mentality was growing within the population. Only about a week prior, John Stoefel had the dubious honor of being the town's first hanging for killing his brother-in-law. Scudder wisely fled back east until tempers calmed before later returning to Denver.

Only two months later, the editor of the *New York Tribune*, Horace Greeley, visited the West to witness the gold rush for himself. While in Denver he said, "I apprehend that there have been, during my two week sojourn, more brawls, more fights, more pistol-shots with criminal intent in this log city of one hundred and fifty dwellings, not three-fourths completed nor two-thirds inhabited, nor one-third fit to be, than in any community of no greater numbers on earth."

The editor of the *New York Tribune*, Horace Greeley.
(Courtesy of the Library of Congress)

Greeley erroneously predicted the crime of the young town would slow in a year, as more people moved into the area. Unfortunately, in the summer of 1860, there were a dozen murders in just Denver alone.

The Duel

As of March 1860 there were three homicides and two duels in the Denver area. One of the duels included the death of Dr. J. S. Stone, who lived in the gold-rich Gregory Mining District. Stone was a member of the Legislative Assembly and sat as a judge in the "miners' courts." His antagonist was L. W. Bliss, acting governor of the territory. During a public dinner the acting governor apparently made an offensive remark directed toward Stone.

A reporter described in *Western Mountaineer* newspaper:

> "After the cloth was removed numerous toasts were drank, and one proposed by Gov. Bliss, reflecting personally on Dr. Stone. The latter instantly arose to his feet and said, 'Gov. Bliss, do you refer to me personally in that toast?' Gov. Bliss emphatically replied, 'I do! Do you wish to see me?'" The public slight was too much for his honor, and so without delay Stone provided Bliss with a formal letter of challenge.

> "Sir:—At the public dinner table to-day, between the hours of one and three o'clock, you used language derogatory to the character of myself. I feel deeply aggrieved at your remarks and demand satisfaction at your hands."

An agent acting on behalf of the governor soon provided Stone with an answer.

> "Mr. Bliss desires me to state that he will be very happy to afford Dr. Stone the satisfaction he desires, on the following terms: at thirty paces, with guns loaded with ball—one or more shots, as circumstances may require. Time, three o'clock P.M., Wednesday the 7th inst., on the banks of Cherry Creek . . ."

The men, their seconds, and a crowd arrived on time for the spectacle.

> "Dr. Stone firing about a second in advance of Gov. Bliss. Dr. Stone (then) fell to the ground supposed to be mortally wounded—the ball of his opponent entering the left thigh, penetrating the bladder, and passing through his entire body. The ball of Dr. Stone struck the ground some ten feet in front of his opponent. Dr. Stone declining a second fire, the parties were then removed from the ground. The whole affair was conducted with much courtesy between the friends of the principals—the spectators were orderly and peaceable, without the slightest demonstration of applause or condemnation, just as though the matter was a familiar, every day occurrence, necessary and proper."

Needless to say, not all violence was served forth in such a refined and eloquent way along Denver's mile high, muddy streets.

The Dangerous Deadline

In 1860 a violent gang set up residence in Denver and called themselves the "Bummers." These men in the early years of the gold rush preferred to make their money off others, and this included a certain familiarity with the use of firearms. Notorious gambler and bar owner Charley Harrison wasn't exactly hesitant about using his pistols on others. He was also connected, by choice or otherwise, to the gang as they used his establishment the Criterion as their base of operations.

Rocky Mountain News founder and editor William Byers was a tireless promoter of the Colorado gold rush. While many came to the area and were discouraged by exaggerated reports of gold, Byers wrote regular articles about where gold was being discovered and how much money could be made in one day. The editor soon found himself in the crosshairs of the gang for an article he'd written. The point of contention came when he detailed a death that occurred at the hands of Harrison, the aforementioned bar owner and gambler. His description of Harrison's character was not flattering; Byers never shied away from printing news that made bad men feel uncomfortable—despite threats that sometimes carried deadly implications. Harrison was no saint and claimed he had killed enough people to create his own jury waiting for him in hell.

Members of the Bummers gang thought the newspaperman had written one too many articles and decided they were going to do something about it. They came to the paper and pointed their guns at Byers. He backed into the newsroom and maneuvered his aggressors until they stood directly under a trapdoor in the ceiling. The gang looked up and saw a number of gun barrels pointing down at them. Byers asked again what they wanted, besides promises to blow out his brains. More calmly they requested his attendance back at the Criterion to speak with Harrison. Byers, no doubt trying to defuse the situation, agreed and let himself be led away to what might have been his death.

Once there, he was taken over to Harrison whom the gang then ordered to prevent the *Rocky Mountain News* editor from escaping.

Harrison, apparently shocked by the group's move, agreed, and took Byers into the backroom to talk, where he admitted to having nothing to do with the gang's actions. One account had Harrison handing Byers a pistol for protection before letting him escape out the back, and another has Harrison and Byers calmly walking down the street and talking nonchalantly as the gang realized their hostage had been freed and threatened to open fire.

Regardless, Harrison was able to impress on Byers the seriousness of the situation and that there would be a gunfight. Byers ran for it, got back to the paper, and found quite a few willing to stay to fight the gang. They armed themselves and barricaded the newsroom. Soon enough the gang began arriving on horseback, including one Jack Merrick. Some harsh words followed by rude gestures had the gang start sending rounds into the little building. Bullets punched through glass, wood, and in some cases shot out clear to the other side. The newsroom responded accordingly, and returned fire.

Merrick was hit with part of a shotgun blast. The buckshot did little damage but convinced him and the others to ride off. Soon word spread that they had killed Byers in the exchange, and the town was furious. The editor was in fact unharmed, but as the gang tried to get out of town many were captured along the way.

Merrick stopped at a girlfriend's house, got patched up, and began to head out of town himself. But something happened. He stopped his horse and turned it around, heading back into town. Whether it was to make sure Byers was finished off or to help rally his own gang is unknown.

Marshal Tom Pollock, who was described as the unusually fearsome combination of a blacksmith and a part-time undertaker, spotted Merrick coming back into town. He grabbed a shotgun, climbed up on his horse, and decided to head him off before he could do any more harm. Both equally armed, the two charged each other and fired their weapons. Merrick's shot missed but the marshal unloaded both barrels of his shotgun, hitting Merrick in the face. The savage blast knocked him clean out of his saddle. Somehow Merrick struggled to his feet, then collapsed onto the dusty ground. He was dead; the newspaper and its editor would live another day.

Three Thousand Miles

After a brutal killing the residents of Denver wanted justice and so proceeded to hire bounty hunter W. H. Middaugh to track the murderer down. They had chosen the right man for the job. Middaugh chased and brought his quarry across three thousand miles, through angry mobs itchy for a hanging and a broken judicial system, to return him to Denver for justice.

It was 1860 and James Gordon was twenty-three years old when he killed an old man. His family had come out west with everyone else during the gold rush but his father saw the future was in ranching, not mining for gold. Gordon himself had a good education and was considered by many to be a polite and amiable young man, which he was until he imbibed alcohol. Even though the book wouldn't be written for another twenty-six years, this was a classic case of Dr. Jekyll and Mr. Hyde. The drink made Gordon nasty, mean, and downright homicidal. It is not difficult to find trouble in a boomtown and even easier to join in. Gordon had begun hanging around with less than desirable types, and things got worse from there.

On July 18 Gordon became drunk and shot a local bartender. The man somehow survived and Gordon sought him out the next day and begged his forgiveness, apologizing for the act. But it wasn't long before his demons got the better of him: just two days later he shot at a man who ran out the back of a saloon where he was drinking. Gordon was angry, bloodthirsty, and wanted to raise some hell. He stumbled out of the Elephant Corral saloon in a red haze, murder on his mind. He spotted a dog, took aim and fired twice, missing, the large caliber slugs thumping dangerously into the dirt by the animal.

Gordon then barged into the Louisiana Saloon, promptly got into an argument with the bartender, and ended up slinging a bottle at an arrangement of other bottles behind the bar. Glass and alcohol exploded in a flurry of shards, and the bar's patrons knew trouble was coming. Many got up from their chairs and ran for the front door. Jacob Gantz, an elderly German, just wasn't fast enough, and Gordon caught the man and flung him to the ground.

Gordon asked the old German if he wanted to share in having a drink with him. Apparently Gantz declined, got to his feet, and made another break to get away. He got outside but still wasn't fast enough. Gordon grabbed him and for the second time flung him to the ground, then he stuck the gun to his head and pulled the trigger. Nothing happened. Gordon tried again, ignoring the pleas of the other man. The hammer fell yet again. Nothing. Gordon pulled the trigger again and again as Gantz continued to beg for his life. One more time the young man pulled the trigger and the gun finally responded, the hammer hitting a live round that sent forth a bullet into Gantz's head. Gordon got up and was triumphant in his murder. He yelled he was glad of what he'd done and wouldn't mind doing it again.

A local gambler told Gordon's family the news and they pleaded with him to save their son from the assembling vigilantes, who were already looking for him. The man found Gordon asleep outside and helped him to escape the mob. Gordon fled and narrowly escaped another group of vigilantes. His only chance was to get as far away from Denver as he could. That's when the town hired W. H. Middaugh to track Gordon down. For three thousand miles he chased the young man, always one step behind. But the day finally came when Gordon was found.

The *Rocky Mountain News* weekly reported that Gordon appeared to be overwhelmed with astonishment when he found himself overtaken and captured.

"The subjoined account of the exciting chase and ultimate capture, and the happy coincidence of circumstances, that enabled the pursuers to accomplish their purpose, furnishes a marked illustration of the truth, that justice, although sometimes slow, is sure to overtake the evil doer in the end. Mr. Middaugh deserves great praise for the promptness and energy he displayed in the fulfillment of his irksome and dangerous mission."

The road back was long, and Middaugh saved Gordon more than once from angry mobs, who wanted to see the murderer dangling from a rope for killing an innocent man. After fighting against an inept judicial system, Middaugh secured the young man from a Kansas jail and returned to Denver. At Gordon's request, Middaugh officiated over his hanging and saw his job through to the end.

CHAPTER 11

NEVER FEARLESS

In the mining towns of Gilpin County it became essential to keep the gunfights out of the streets and to keep the criminals away or dancing at the end of a rope. As gold mining moved from the streams to the mountains, it became quite clear that investors would be needed to continue to remove the gold from deep within the rocks as the process was costly in both men and equipment. In quick order both law enforcement and courts were devised to try to prevent the camps and later the towns from descending into lawlessness. And for the most part, it worked.

William "Billy" Cozens came out west with the gold rush to get rich but soon found the reality of gold mining difficult and that it offered more busts than bonanzas. He turned from bartending to law enforcement, something that maybe wasn't that far of a stretch as they both required sound judgment and the ability to let the steam out of a tricky situation.

Jack Keheler was the first elected sheriff of the Gregory Mining District and Cozens became his deputy. There was no jail, and not much to work with. Cozens said he had two Colt revolvers, a Sharp's rifle, and a pair of handcuffs and leg irons. That's it, and he just had to be swift. Being a mining district, its residents were summoned whenever a miner's court needed to be convened to dispense justice when a crime was committed. Sentencing was a relatively straightforward process as more severe crimes resulted in death. This kept things simple and demonstrated to would-be criminals that there would be sparse leniency.

Acts of murder resulted in being hung—but years later Cozens said it was common to hang someone who stole gold out of a sluice box or stole a horse. In 1862 Cozens became the elected sheriff, a position he held, more or less, for two terms. Cozens also served as a US Marshal, commanded two companies of infantry during the Indian Wars, and was considered by and large to be fearless—which, he readily admitted, he wasn't remotely.

Cozens said the most scared he'd ever been was just a few months after being made sheriff. A man had shot another man and ran to hide in an old mining shaft. By the time Cozens got to the mine, a large crowd had already formed outside ready to string him up. As the new sheriff, Cozens soon came to realize that the rule of law in the mining district was a tenuous one. He knew that he'd have to go into the mine and bring the armed criminal out. He was handed a candle for light, and so equipped, he entered into the dark.

Sheriff Cozens went inside and descended the ladder, the flickering candle no doubt providing little in the way of light. As he made his way down he heard the sound of a pistol's hammer being cocked. The click always preceded a weapon being fired. Thinking fast, Cozens dropped the candle and swung to the other side of the ladder. He knew the man had seen him before he dropped the candle, and so knew where he was on the ladder. He figured he was in a unique spot of trouble.

Cozens began talking to the shooter, who was at first silent. The sheriff told him that there was a crowd outside waiting to kill him and that his job, as the local sheriff, was to see that he got a fair trial. All the while he dangled from a wooden ladder in a dark mine shaft as a murderer pointed his gun up at him. But Cozens was convincing. He climbed back to the top and dispersed the crowd with the exception of about twenty men, who vowed to see the man hang. Cozens pulled out his revolvers and told the would-be vigilantes to leave before he began taking aim. It worked: the men left, the murderer got his trial, and he was convicted. And like all successful convictions from this time, he was hung.

The area's first jail came into being after Cozens caught two horse thieves and needed someplace to put them until morning when court could be held. Cozens decided, quite incorrectly, that it would be a good idea to take them to his house and handcuff them to the foot of his bed where his wife was sleeping not far from their newborn baby. Cozens recalled that his wife was, as one can imagine, somewhat unhappy at the prospect but relented. Cozens warned the men if they disturbed his wife he would kill them on the spot. The next day his wife told him, in no uncertain terms, he was not to use their bedroom again as a jail cell. He ended up getting permission from the county commissioners to use prisoner labor to build a jail. As there were no prisoners, he went around and gathered up "volunteers" from the area's less than savory populations, and soon the jail was finished.

On one occasion, he arrested two men guilty of robbing sluice boxes in the town of Black Hawk, Colorado, and recalled they asked to see a lawyer. Their lawyer then asked Cozens where his warrant was for their arrest. This was a new question local law hadn't faced before, and so after a moment of thought, he took out his two pistols and put them on the table. He then informed the attorney that those were his warrants for each man. This seemed to put an immediate end to the issue.

In the summer of 1865 a group of bandits worked the Gilpin area, holding up horse-drawn coaches or stealing shipments of gold from some of the more isolated mines. Members of the group were reputed to wear two six-shooters each. The Bobtail Mining Company put its gold bullion in a safe before shipping it. The men bribed a miner for details and planned to make a raid on the mining company. The miner, having an attack of conscience, told Sheriff Cozens about the bandits' plan and when they were going to make their move.

Cozens and several other men got to the location first and waited for the bandits to arrive. When they did, gunfire broke out and three of the four bandits were killed outright. The fourth was left wounded on the ground. The next day people came to the mine to see what had happened and found the bodies were left where they fell. The wounded bandit was still bleeding and cursed anyone who came too close. The man's guns were taken away, and Cozens protected the bandit from the people eager to lynch him. None of the local doctors wanted to treat the man, who soon died from his injuries.

Despite the area's best efforts violence and crime did find its way into Gilpin County, but not at the rate of other Colorado towns. This was due in part to the serious efforts taken by police and the towns to slap on a veneer of respectability.

One such attempt to bring culture into the area included the theater and opera. In July 1862 George Harrison, owner of the National Theater, was tried and found not guilty after he fired thirty-five shots into rival saloon and variety hall owner Charlie Switz at the Concert Hall. Another account has Harrison emptying both barrels of a double-barreled shotgun into his adversary at the entrance of the Barnes and Jones' saloon. It's said Switz was not liked, and as such the jury found Harrison not guilty.

A Phantom

Not everything on the business end of a bullet was alive when it got there. Newspapers from that time had accounts of phantoms troubling the living such as the ghost of Bainbridge in Georgetown. Central City had several similar accounts but only one where half the citizens that came across it also tried to kill the ghostly specter.

"For three weeks past, several of our citizens have carried in their pent-up bosoms a great and fearful mystery. It has come out at last. We have a verifiable ghost in our city," a reporter wrote on the front page of January 19, 1874's *Weekly Register Call*. "He, she, or it, appears on Eureka Street, near the mouth of Prosser Gulch. This portion of our city was once a scene of great activity. For years the mills have been abandoned, and most of the former dwellings, stables, etc. have from disuse, fallen into a state of dilapidation, calculated to invite the occupancy of unrestful spirits."

The newspaper described how the spirit often showed itself to fearful residents between 9:00 P.M. and midnight.

"Being an independent spirit it declines to conform to the usual ghostly fashion, but dresses to suit the season and altitude. Sometimes it appears in long drapery of a dark color; sometimes its robe appears one side white and the other colored; it wears a cap, and sometimes boots, with a suspicion of pants, besides exhibiting other idiosyncrasies."

The article said numerous local witnesses saw the apparition and one "respectable gentleman" spotted it twice near the steps to his front door.

"The appearance seeming to arise from another flight of steps, passing across the sidewalk before him into the street. The first time no word was spoken. The second time the gentleman asked him who he was, and what he wanted, and remarked that if he appeared to him a third time he would get a ball through him. This elicited no response, but it is plain that his ghostship does not wish his clothes mussed, as he had let this gentleman carry his navy (pistol) for nothing ever since."

Apparently, other members of the town's citizenship did get the chance to shoot at the phantom as it made its way about the town.

"On one occasion the apparition chased two men from the neighborhood of Newell's lumber yard down to Mack's house, when he disappeared in the tunnel, perhaps. Returning immediately, they again found him at the lumber yard. One man who had been a number of times confronted by his ghostship, concluded to put a stop to his (visits) . . . and sent, as he believes, a couple of pistol balls through him at different times, but with no other effect than that he vanished, faded out of sight like the thin fabric of a vision."

Another resident wanted a crack at the ghost, and when he arrived in an area known to be haunted, he waited for it to show up, and it did. The form raced up a street, its robes flowing behind it. The man took careful aim at the spirit, prepared to squeeze the trigger, and then stopped at the last second.

"He discovered that this was anything but a ghost, [it instead] being the portly form of a well-known citizen who had been making a call in that neighborhood, and was returning home with a proud, elated tread."

Rather than the much more sensible argument to refrain from shooting at presumed phantoms, the newspaper then took the tack of trying to persuade readers not to wear cloaks in that area at night or to whistle a jaunty tune lest they be shot. Speculation then arose as to who the ghost might be, with the newspaper considering a host of prime suspects who died or were killed in a number of grisly ways.

"Who it is remains a mystery, but . . . we give him the kindly warning to buckle on a bullet proof armor or retire from Eureka Street."

Remnants of the Past

Between the deaths caused by bullets, mining-related accidents, and any and all things associated with living more than 150 years ago, the cemeteries soon grew quite full. Central City's cemetery was at capacity before both the need for housing and the discovery of more gold forced its relocation.

"This museum is on the original cemetery," David Forsyth, executive director and curator of the Gilpin Historical Society, informed me as we sat in his office. I glanced outside and saw that like much of the city, it was located on a near-vertical hillside, old houses and buildings built almost on top of each other. The road to the museum and my appointment with Forsyth was unforgiving and narrow, giving me the impression that this was a town designed by someone who apparently didn't suffer fools or those with second-rate GPS phone apps.

"This originally was not desirable land. There weren't any mining claims . . . but eventually, they did start mining here," Forsyth said. At first called the Winnebago Hill Cemetery, the decision was made that there just wasn't any more room for the escalating number of bodies.

"In the mid-1860s it was getting full up here, and they needed more housing, so somebody built a house up here while it was still the cemetery. More and more houses were coming in, and they said, 'All right, we got to move the bodies.' So that's when they started City Cemetery at the top of Eureka Street."

There were some setbacks to moving the graves, namely that they were in the wrong places, not marked, or missing headstones.

"They did the best they could," Forsyth said. "They do still occasionally find bodies. The last time it happened was about five years ago when the owners of the house next door found three bodies."

He also mentioned an incident when the sewer line broke, and they found old bones and shoes. The contractor decided digging was a bad idea and opted to drill. "Then they hit a coffin," Forsyth said.

Then there was the time in the 1970s when a young boy was raking in the front of his house.

"And he comes running in, [saying] 'Mommy, Mommy, I found a coconut,'" Forsyth said, adding it wasn't a coconut. "It was a skull with hair

The Knights of Pythias Cemetery, one of several cemeteries located in Central City. *(Photo by Chancey Bush)*

still attached. There was another woman who was doing a flower garden and unearthed a body. So it still happens."

He said there are worse things than accidentally unearthing a forgotten grave from the early days of the gold rush. "We've got a sinkhole outside of the museum here, and we're all kind of hoping that it is just a grave and not a mine shaft."

That's because a grave leads to a body, but sometimes no one knows where a forgotten mine will lead or how deep. And while bodies rest, mines never do.

When at last the law arrived in the towns struggling in the shadows and valleys of the Rockies, it did so in fits, and it took time to stick. From the mountains to the plains, there were always those who believed payment in lead was just as good as gold. But in the West there were things worse than just dying.

CHAPTER 12

WENDIGO

The Algonquin believe there exists a dangerous race of cannibals who hunt and kill humans that enter into the forest. They are called the Wendigo. When the snow falls deep and food is hard to find, and winter crawls across the land, these monsters, with their hearts of ice, attack and kill anyone they find. And then eat them. Their mouths are twisted, and they are known by their frightening voices and high-pitched whistles in the night.

As Thursday morning came, the darkness of night shrank back, becoming a jagged line of shadows along the snow-shrouded forest tree line. A bit of the inky blackness tore itself free and came shambling through the snow to the Los Piños Indian Agency in Saguache County. Several officials stationed at the government outpost were sitting down to breakfast when a wild-looking man was spotted coming toward them. He was ragged, his long black hair and beard matted. His eyes were fixed on the agency, and with rags instead of shoes, he continued trudging forward. His desperate journey would take him into a unique and utterly horrific historical footnote in the search for Colorado gold. It was April 16, 1874, and Alfred Packer had once again found civilization.

The men ushered their strange guest inside, out of the cold, and offered him food. Some reports stated that despite being stranded for sixty days in the unforgiving wilderness he had no taste for meat, while others said he only requested a glass of whiskey. Whatever the truth, the impression the thirty-one-year-old left on them was no doubt an intense and troubling one. Packer was described as being muscularly built and pale with black hair. Sometimes described as having wild eyes, his most prominent feature was his unique, high-pitched voice, which some found both agitating and distressing. To top it off, it's said he also often referred to himself in the third person.

Packer sat down and told them a story of how it came to be that the small party he traveled with were not with him anymore. This was only one of several stories he would tell about what happened in the wilderness the previous February. A story about how his five companions left him on his own because he was snow-blind and footsore. He said he didn't know where they were. He was, of course, lying. Because the truth was, he'd eaten them.

The Fortune Hunters

For whatever reason, Alfred Packer was not fond of his first name and often went as Alferd. He liked the name so much more than his slightly different given name that he had it tattooed on his arm. Born in Pennsylvania in 1842, Packer likely had a problematic childhood. Many years later he often wrote disturbing and threatening letters from prison to his surviving relatives about his difficult upbringing. While incarcerated, he promised to kill his sister after he was released. Packer suffered from epilepsy, which proved something of a curse most of his life. He had enlisted in the Union Army in 1862 to fight the Confederacy but was discharged not once but twice for his medical condition. The discovery of bromide to help control seizures enabled him to hold a host of odd jobs, including miner and hunter. Legend has it that in Georgetown, Colorado, not twenty miles from where George Jackson had found gold, a mining accident claimed parts of fingers on Packer's left hand.

In the fall of 1873 a discovery of gold in Breckenridge, Colorado, grabbed the attention of a group of fortune hunters in Provo, Utah. Gathered in a boardinghouse the party of some twenty enthusiastic would-be prospectors came together to plan and set off on a journey into the mountains.

Then Packer arrived, caught wind of their plan, and was all too happy to tell the group that he was an experienced Colorado gold miner and guide. For a grubstake in their expedition, he'd show them where to find the gold. With gold fever running high and impairing better judgment, they set out for the snowy Colorado wilderness in November. By January the harsh weather took a vicious toll. They were out of food, and if not for the intervention of a band of Utes, their situation could have taken an early sinister turn. That would come later.

The Native Americans took the group to Montrose, and Chief Ouray offered to feed and shelter them for the duration of the winter's worst, advising against their current plans. But a discovery of gold is as much a lure to unimaginable wealth as it is a desperate, dangerous illusion. Claims are staked out as fast as they can be discovered, regardless of the wealth hidden within the rock and gravel. Those who missed out often went home empty-handed and much worse off than when they had arrived. They were gambling with their fortunes and lives, and the promise of gold had always proven a dangerous one.

A small party within the group tired of waiting and broke off from the rest to brave the winter storms. They were almost immediately blasted by a severe blizzard, became hopelessly lost, and almost died. So hungry did they become once their food ran out that they killed with their bare hands a government-owned cow they came across, eating the meat raw. Fortune favored them, and they soon encountered a cattle superintendent who took them in. But gold fever was already spreading among the men, and no doubt images of their companions finding and pocketing gold spread across the fertile fields of their imaginations like a prairie fire.

A second small party from the original group could wait no longer and set off for the goldfields on February 6. This six-person group included Alfred Packer. A handful of weapons among the party included a small number of firearms, a knife, and a hatchet. Packer was unarmed, but he wouldn't remain that way long.

The six men were seen leaving the Uncompahgre Valley as another winter storm began to build on the horizon. The trip was estimated to take one week, and as such, they brought just enough provisions for that span of time.

Two Months Later

The specifics of how exactly the party lost its way are wiped from pages of history, the details smudged by the always contradictory and sometimes rambling tales Packer spun in the air before his questioners. In the first version of the story, his companions left him behind. But with great luck and tenacity, he safely made his way to the Indian Agency. The government officials there noted that Packer had recovered very quickly for a man on his own in the unforgiving Colorado wilderness for two months.

Claiming to be poor, Packer sold the rifle he had brought with him for $10 and was soon ready to continue traveling. He left the Indian Agency and the questions of his missing companions behind for Saguache. The lure of gold still glowed over the mountains in Breckenridge, and Packer heeded its call. Once in Saguache, Packer became a big spender, having much more money than he had originally admitted to. By then all the original, and perhaps wiser, members of the gold-hunting party had left the hospitality of Chief Ouray and discovered Packer in the settlement.

The mere fact that he gambled, drank, and was seen with money he didn't have previously made the others rightfully suspicious, not to mention the other five men still hadn't arrived. Suspicions grew, and Packer was being questioned again by authorities. This time, his story included the party leaving him with the rifle he later sold. It was with this weapon, he said, that he was able to kill and eat a rabbit, which helped him to survive. Maybe it was his unusual demeanor, changing stories, or unexplainable wealth, but they weren't buying what Alfred Packer was selling.

He was asked to return to Los Piños to answer further questions, and for whatever reason, Packer agreed. It was during the trip that they ran into Jean "Frenchy" Cabazon from France. He was also part of the original party that had stayed behind with the Utes. Apparently, Cabazon had heard the tale Packer had spun of his misadventures and eventual departure from his party. Cabazon flat-out said he didn't believe Packer. Then Packer called the man a liar in his high-pitched voice and promised to kill him the next time the two met. The two parties went their separate ways, and as fate would have it they would indeed meet one more time.

Back at the Indian Agency, Packer was again questioned about the events that led up to his being left behind. But Packer's story had holes and wild inconsistencies that were easy for those questioning him to use to tear apart his account. And, with what would later become a recurring trend, Packer revised his story with yet another confession of the events that transpired. This tale was one of systematic, cannibalistic attrition.

In this story, the men left the Ute camp and were soon lost in the frozen hell brought on by a blizzard. Their food ran out and the oldest man in the party, sixty-year-old Israel Swan of Missouri, eventually died of starvation. The men then decided it was the best idea for their mutual survival to cut Swan into pieces and eat him. With their bags packed full of life-sustaining, if not palatable, Missourian, they headed out yet again. But before too much time had passed, James Humphrey of Philadelphia died. The men decided that eating him was an excellent idea as well.

Packer said it wasn't many days later when redheaded Shannon Wilson Bell of Michigan shot George Noon from San Francisco. Necessity forced them to eat the German butcher Frank Miller and, perhaps as no surprise, Noon as well. Packer said Bell then tried to kill him with the butt of his rifle one day while they sat around the campfire. Packer moved aside and killed Bell—then ate him too.

Apparently, his interrogators were sickened by Packer's story of cannibalism made even worse by his recounting of the gory details in his trademark shrill voice. He swore this latest version was the truth and even signed it before the local authorities. But those who listened to the grisly tale again doubted the truth of Packer's account. And they were right to.

The authorities asked Packer to lead them into the wilderness in search of the bodies of the five men. With Packer leading the way, the search party left for the Lake Fork of the Gunnison River. When they were nearing the area, they found a pillbox with Packer's name on it, but they could not find the bodies of his companions. They even drained a beaver dam to see if the bodies were hidden beneath the water. But try as they might, the dead men could not be found. For his part, Packer said he grew confused and couldn't remember where he had left the other five men, nor could he provide any useful details to the search party.

After some unsuccessful attempts to find the bodies, Packer was arrested on the suspicion of killing his companions. But the truth was he knew exactly where their bodies were and almost a decade later claimed

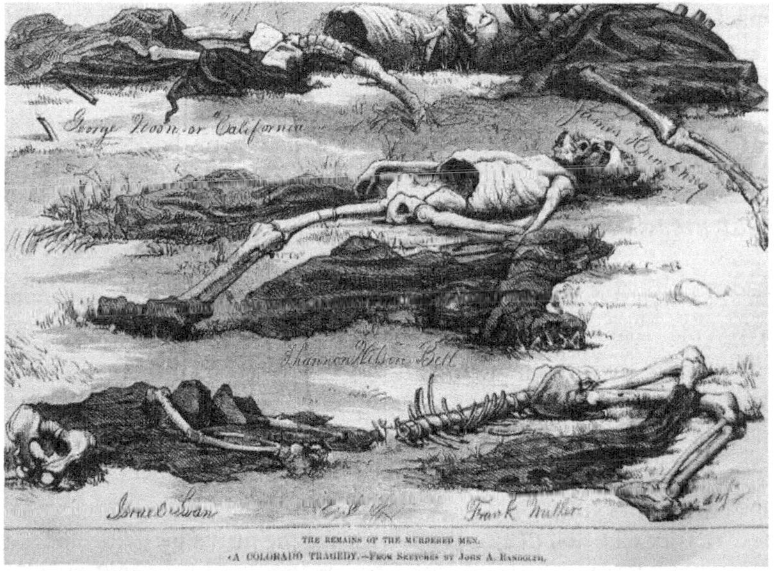

Artist John J. Randolph came across and illustrated the remains of the men who traveled with Alfred Packer in 1874. *(from* Harper's Weekly, *October 17, 1874)*

that "they" told him not to bring the searchers to the spot. Of course, he never fully explained who "they" were. Perhaps wisely the authorities decided to arrest Packer, with or without the bodies, as they suspected something wasn't quite right. He was placed in a makeshift jail and it wasn't long before the bodies were indeed discovered. While there's some disagreement about who first discovered them, artist John J. Randolph came across the remains in August, not far from where the search party had ended their efforts.

The illustration he drew of the scene showed all the men in various poses and states of decomposition among the grass. Skeletal legs gave way to gaping ribcages and in some cases were topped with completely intact, bearded human faces. As far as it could be determined, there wasn't much of a struggle at the campsite. One man was said to have had part of a blanket buried into his skull from a hatchet attack. Another's head was missing, likely due to an animal, and wasn't found until much later.

A warrant was ordered for Packer's arrest. He was wanted by the authorities dead or alive. But in what would be just one in a long series of mind-boggling coincidences, Packer escaped from the jail before the warrant could be served. He later claimed that his flight was made possible by an unnamed accomplice who passed him a key crafted from the blade of a knife. Free from the jail, Packer slipped away and wasn't seen again for nine years.

Hatchet to the Head

In 1883 Jean Cabazon stopped in a roadhouse on his way to Cheyenne, Wyoming. That evening, while getting ready to turn in, he suddenly heard through the walls the unforgettable voice of Alfred Packer. His was the kind of voice you only had to hear once, and no doubt Cabazon remembered Packer's passionate promise to kill him if they met again. Cabazon decided to take the chance and meet with the owner of the voice the next day to confirm his suspicions.

They met, and there was no mistaking the man, now going under a false name. For those nine years, Alfred Packer lived in several western states doing his best to avoid being discovered by the authorities. Apparently, Packer did not remember Cabazon, so he took his first opportunity to alert the local sheriff of Packer's real identity. Packer was then arrested for a second time, without drama, and delivered into the hands of the Colorado authorities. As he was questioned again about what happened to the doomed party, his final tale predictably changed and became even more gruesome.

Packer described how the small party left Ouray's camp to head for the gold discovery in Breckenridge. As in past stories he explained how they battled the savage winter weather until they ran out of food and became exhausted beyond belief. No matter how they tried, they could not find anything to eat. Packer said under the pressure of hunger and imminent threat of death, many of the men broke down in tears. He said Swan soon became crazy from the never-ending series of hardships.

One day Packer was asked if he could see anything from a nearby mountain to help make their way out of the wilderness. Taking the rifle with him, in case he came across any game, he began his attempt to see out of the area they had found themselves lost in. It turns out he couldn't find any food, or any way out of their predicament, and so decided to head back to the party's small camp. The day had nearly ended before Packer was able to return.

"When I came back to camp after being gone nearly all day I found the redheaded man [Bell] who acted crazy in the morning sitting near the fire roasting a piece of meat which he had cut out of the leg of the German butcher [Miller]," Packer recalled later in an unusually coherent letter to the *Rocky Mountain News,* including the little touch of irony in mentioning Miller's profession. "[Miller's] body was lying the furthest off from the fire down the stream; his skull was crushed in with the hatchet. The other three men were lying near the fire; they were cut in the forehead with the hatchet. Some had two, some three cuts."

Packer said when Wilson saw him arrive into camp, he charged with a hatchet. Moving fast, Packer shot him in the stomach, and Wilson fell forward onto his face. As the hatchet slipped from his grasp, Packer said he snatched it up and hit Wilson on top of the head, killing him. All the men except Packer were now dead due to fatal hatchet strikes to the tops of their heads. This is the story Packer stuck with until the end of his days.

Surviving

Packer said after killing Wilson, he tried again to escape the area and failed. The snow was too severe, and he had to return to the camp of dead men. Starting a new fire, he looked around and saw the partially cooked portion of the German butcher and decided to continue cooking it. He told authorities that eating that piece of meat made him sick, and yet as he tried to escape for the next sixty days, he ended up eating all the others as well.

"I tried to get away every day but could not so I lived off the flesh of these men."

One day, Packer finally escaped, and using the last pieces of human-based sustenance, he built up his strength and was able to reach the agency.

Packer's first criminal trial was in Lake City in 1883. He told the members of the jury how desperate he and his party were, how they took turns anteing up their shoes for their rare meals. He declined representation by a lawyer and chose to conduct his own defense. In this way, the jury could listen to the man speak in his unusual voice sometimes for hours on end. Seemingly oblivious to his circumstances, he also threatened to kill a witness for telling lies about him.

His story of what happened changed very little this time around, but the prosecution mentioned that Wilson was likely shot in the back and not in the stomach. When the time came, the jury returned with their verdict in a quick three hours.

The judge eloquently sentenced Packer to death by hanging. But Packer would show, not for the first time, to have the devil's own luck. A technical hiccup in the law saw that his sentence of death was removed because the state had repealed its 1870 murder statute. And in Packer's favor, the state hadn't yet agreed on a replacement statute. It was decided Packer would need to be retried. He was transported to the jail in Gunnison because many believed he'd be safer from would-be lynchers looking to dispense some traditional frontier justice for the now notorious man-eater.

His trial this time would be for manslaughter instead of for murder. Packer stayed in the jail for three years, and during that time the county's sheriff believed Packer to be dangerous. In addition to threatening letters sent to Packer's family members, the sheriff thought Packer was particularly vicious and completely lacking in redeeming features. The lawman also said he learned, during Packer's incarceration, details about how his prisoner had killed two men after his escape from jail near Colorado Springs so he could steal their supplies. The sheriff also believed Packer killed a prospector in Arizona. Packer apparently confessed to these additional murders during a series of conversations.

In 1886 Packer's second colorful trial began, and he was convicted of five counts of manslaughter for a total of forty years in prison. He insisted that he was guilty of killing only Wilson, but the jury was inclined to disagree. When looking back at the facts of the case, Packer's escape from justice, repeated lies, or half-truths, might have caused the jury to disregard any version of the events that he told. Not to mention the fact that the man he admitted to killing had ultimately died from a hatchet blow to the top of the head—exactly like the other four—an unlikely coincidence.

During Packer's incarceration after the second trial, it is said that he was well behaved, a perfect prisoner. In time he even garnered a sympathetic following from people who believed he was over-sentenced for something that could never possibly be proven. After all, it is easier to believe he was a man in the wrong place at the wrong time and a victim of circumstantial evidence than to believe he was a man who singlehandedly killed his five companions. Behind bars and sentenced to heavy labor, he was unable to cause any more harm. Of course, that didn't stop others from causing harm as a result of Alfred Packer.

His plight drew the attention of the *Denver Post* newspaper and specifically reporter Leonel Ross O'Bryan, who wrote under the amusing moniker Polly Pry. Her editorials decried the injustice of Packer's conviction, and she even started a petition that netted a hundred signatures, many from notable members of society. One day attorney W. W. Anderson approached O'Bryan and said he'd discovered a cunning plan to get Packer released from prison that had to do with the crime happening on a Native American reservation. This revelation added further hiccups to the charges Packer faced and wasn't an idea that was completely new to the editors of the paper.

However, the *Post*'s owners, Harry Tammen and Frederick G. Bonfils, met with the attorney and entertained his plan to free Packer. They added in no uncertain terms that they wanted to consult the paper's own lawyers before proceeding. Despite their request to wait, Anderson went to Packer and falsely claimed to be with the paper and was able to retain Packer's power of attorney. Packer also paid Anderson a modest sum of the money he received from the military for his brief enlistment during the Civil War.

The paper's leadership soon learned of the false representation and grew furious at Anderson's indiscretion and betrayal of their trust. They demanded to meet with Anderson after hearing about the stunt he pulled.

After a loud and heated conversation, Anderson was physically thrown from the office. He pulled out a concealed pistol and yanked open the office door, firing at the two inside. Bullets hit both of the newspapermen. As Anderson took careful aim for a fifth shot, O'Bryan, who happened to be in the room, confronted him and knocked his gun aside. Anderson fled. Both of the paper's owners recovered in time. And ironically, Packer's would-be attorney got off the hook when his trial ended, incomprehensibly, in a hung jury.

In 1901 on the day before being succeeded by the newly elected governor, sitting Colorado governor Charles S. Thomas permitted the notorious man-eater parole. He had long been aware of Packer's threatening letters to a family member but released Packer due to his failing health on the condition that he never again leave the state. Colorado would become his new prison. Packer was jailed for seventeen years before once again becoming a free man. It is said he lived the rest of his life quietly and alone. In his later years, he was well known among the local children of the community where he lived and told them exciting tales of the Old West.

But likely not his own—which was one of the most colorful and bloody to come out of the state's desperate search for gold. He suffered a serious epileptic seizure in 1907, and after several months of being bedridden Alfred Packer died, taking the truth of 1874 with him back into the dark.

His story shows that people searching for gold will put up with hardship, perilous uncertainty, starvation, and worse. The dangers involved with prospecting and mining didn't disappear with the Old West, and some, it could be said, thrive on it.

CHAPTER 13

THOSE WHO CAME BEFORE

Fueled by greed and a desire for conquest, the Spanish came to the West looking for the mythical cities of gold hundreds of years before Colorado's gold rush. Sixteenth-century explorer Álvar Núñez Cabeza de Vaca returned to Mexico City after years of exploring the Southwest with captivating stories of secret cities of unbelievable wealth. These tales were told to him by the native populations he had met during his many travels. He had heard stories of the Seven Cities of Gold, said to have roads paved with gold and where its people ate off gold plates.

While the possibility of finding abundant golden dinnerware among the indigenous peoples of the Southwest was extremely unlikely, the Spanish were intrigued.

In 1538 a priest named Fray Marcos de Niza was given orders by the local government to go into the new lands to the west to help provide further accounts of these legendary cities. He and his small group, including a Moorish slave named Esteban, set off to find Cibola, one of the Seven Cities of Gold. Along their journey Esteban traveled ahead of the group, and it is said he was killed by a Native American group he came across. Fray Marcos de Niza heard the news, but decided to forge onward to continue his search, going ever farther into the new lands. In Arizona, he saw a large city in the distance that seemed to shimmer in the hot air.

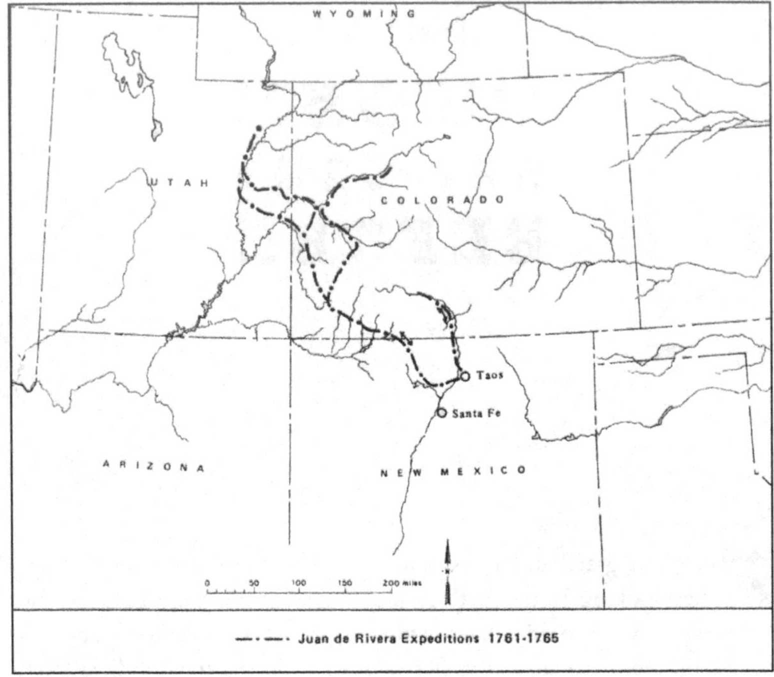

Early Spanish expeditions into Colorado. *(Courtesy of the Bureau of Land Management)*

Believing it was surely one of the seven cities, he returned to Mexico to tell his story. The Spanish Crown, ready to continue its expansion after conquering Mexico and Peru and with tantalizing stories from Álvar Núñez Cabeza de Vaca and Fray Marcos de Niza, organized an expedition to go north.

With the goal of expanding their empire, Spain sent conquistador Francisco Vasquez de Coronado with a host of Spanish soldiers and natives in 1540. The soldiers marched in their armor across the dusty southwest and left behind them a path of destruction. Their native guide led them through what is now southeastern Colorado into Kansas. However, they found no golden cities. The guide, nicknamed "the Turk," was later strangled and killed by Coronado's men who believed he had intentionally lied to and misled them.

While no cities of gold were discovered, the Spanish became interested in precious minerals. In the early 1700s, the Spanish government provided royal grants for mining. The empire still longed to find wealth in the north. An expedition was sent into Colorado to search out and find gold and silver. Explorer Juan María Antonio de Rivera took his men from Santa Fe into Durango and Gunnison and spent the time from 1761 to 1765 in the Colorado Rockies. When he returned, he brought with him samples of gold ore—but the Spanish government, thinking it wasn't enough, ultimately chose not to continue their search. But there was an unimaginable amount of gold still to be found in the Rockies. When the first prospectors came into the mountains, they found traces left by the Spanish from a hundred years before.

Idaho Springs miner and author Merle Sowell theorized that those early American prospectors and miners often just picked up their shovels and picks and began to dig in the remnants of mine workings the Spanish had left behind.

Not for Money

Jesse Peterson reached deep into his pocket and took out a gold nugget the size of a small orange. Casually handing it to me to inspect, I was surprised by its weight. Over the last few years, Peterson has worked to open the Chaffee Mine in Gilpin County.

"Little by little we've been doing something over here," he said, motioning over his shoulder to the yawning mine entrance behind him.

Peterson, a former Marine, has long gray hair and beard, work clothes faded to an indiscriminate color, and a camo-colored hat. He has worked the same plot of land for forty years. When he began his recent work on the Chaffee Mine, he discovered remnants of the Spanish.

"We dug this all out because all the timber had collapsed," Peterson said of the mine. "While we were digging we found little beads of gold and silver and lead. When you get down into the dirt, we'd find burned firewood, burned tree branches." Peterson added that when the Spanish would work an area, one method they used was to find an outcropping of ore and set fire to the spot.

"They'd burn all the sulfur out of there and in turn maybe melt all the gold and silver down, then they'd go through the ashes, take it, and process it," Peterson explained. "The Spaniards pretty much did everything in the front of the mine because they didn't have a lot of heavy explosives. They probably used black powder but it was fairly dangerous. If they ran out of slaves, if it wasn't easy for them to take it out, it was kind of like a flash in the pan for them."

Peterson said the old-timers, or prospectors from the late 1800s, would instead find a vein of gold and follow it deep into the mountain. If indeed the Spanish had mined the area, then Peterson is just the most recent miner looking for gold on the property over the last 250 years.

"I've had it forty-one years. In the beginning I didn't do much but just the last couple of years I've been doing a lot."

It was a challenge that brought him back from the edge of a serious illness that had him bedridden. The work was difficult, and always dangerous, and it started with freeing the mine of the layers of ice that had filled it up over the years—with a chainsaw.

Claim Jumper

Originally from Milwaukee, Wisconsin, when Peterson got out of the military in the '70s he ended up working in a factory that built mining equipment. While the plant was shut down for two weeks to make equipment adjustments, he and his wife decided to come out to Colorado for a vacation.

The inspiration came after reading 1867's *The Mines of Colorado,* by Ovando J. Hollister. Peterson decided to come to the Centennial State and try his hand at doing a little prospecting. He spent the first ten days panning along North Clear Creek: "Sat down and from morning till dusk digging gold, running a sluice box, panning the gold in front of a headlight."

One day he spotted the property's owner, "Vic," coming down a steep hill carrying under each arm two potato sacks full of gold ore.

"He comes down, and I'm asking him about gold panning, and he says, 'Well, let me catch my breath' and we started bullshitting," Peterson said. "And he told me he was going up there to get (ore) concentrate to pan because it is full of gold." He asked Vic if what he had collected in his own pan was gold.

"'That's mica, and the first thing is you need to burn the [new] gold pan out—because there's oil in it.' He said, 'If you want to learn how to pan gold I'll charge you $10 a day. You can camp out on the property and I'll show you how to do it.'"

And so Peterson began to learn the tricks of the trade from Vic alongside the stream. At the end of ten days, he'd found about an ounce and a quarter of gold, which was then worth only about $100.

As Peterson told his story his son, Jesse, and his grandson sat with him at the front of the mine's entrance while a dog looked from face-to-face for attention and occasionally whined.

Peterson said he found his future one evening as he and Vic were sitting around a campfire, drinking Jack Daniels and panning by the dim light.

"He told me he was going to sell the place, and I said, 'OK, well, I'm going to buy it,'" Peterson said. "Come to find out that he was thinking of selling it to me, figuring that I wouldn't succeed, and then he could sell it to someone else again. He didn't realize that I wasn't going to give up." But there was a catch. For years, Vic and another man were at each other's throats over the piece of property.

"You see, Vic had a claim jumper that was trying to take this claim," Peterson explained. The two had taken turns burning down the other's shack and things were just beginning to heat up. So Peterson decided to solve the problem with the claim jumper. He said he did so in two ways. The first was through legal action; the second was more direct.

"One time he reached into my vehicle to grab me," Peterson said. "I rolled the window up and drove him up and down the drive. I said, 'You about done yet?'" He added it took a couple passes up and down the road before the man finally relented. The issue, he said, was eventually settled in court.

High-Grading

As the weeks went by, the litigation was settled, and Vic came across a pocket of placer gold on the property. Unfortunately, Peterson didn't know how to use the large equipment to recover it. He made a deal with

Vic to split the gold 50/50 if he could help him with the task. He did, and they removed $70,000 in gold, which Peterson used to pay off the property.

One day, while watching Vic pan in the river, he saw the older man get up and go to the front seat of a dump truck and come back again. He did this several times, and the act became suspicious. Peterson walked to the front seat of the truck to take a look around and found in the ashtray a cellophane wrapper full of gold nuggets that Vic was secretly stashing away.

"So I go, 'Oh, you're high-grading me?' Soon the equipment broke, the partnership came to an end, and the two parted ways. "It's all part of it. I don't hold it against him. He's a high-grader, and that's what a miner does. He got his pay for the place, and I got my money to keep going and I didn't have a claim jumper around."

Peterson began upgrading the operation. At first he ran ten yards of dirt each day, filtering out the gold. Then Peterson moved on to larger equipment that could handle sixty to one hundred yards an hour. He got to the point where he was removing three ounces of gold per day from his placer operation. He didn't really pay much heed to the Chaffee Mine on the property and soon let tourists come to receive panning demonstrations and to try their own luck. His son grew up on the property, gold panning and teaching tourists how to find gold. They both remember working an underground placer deposit one December and literally just picking up chunks of gold off the bedrock.

"That's why I don't get the 'gold fever.' It's like, I did it when it wasn't popular and it was a whole lot of fun," Peterson said.

Downsized

Central City and Blackhawk, two mountain towns located just west of his operation, established legalized gambling in the early '90s, and with it came increased scrutiny of his operation from the state government.

"It started getting to the point where it started not to become fun anymore," Peterson said. "When I first started in the '70s we went out to have fun, we didn't go out to get rich. When it stops being fun, you need to do something about it."

He said they took large portions of his land and wanted him to pay for a $60,000 stormwater survey.

"I said, 'Forget it.' I just let it go and downsized," Peterson said.

Then he started having kidney issues and got terribly sick, so ill he couldn't get out of bed. A geologist friend of his asked him one day if he could take a look around inside the Chaffee Mine. Peterson agreed but said he wasn't feeling well enough to go inside with him. It had been many years since he'd dug inside the mine and it had started to fall down in places and become unsafe from the lack of regular attention. His friend came back to him later that afternoon and told him the mine had an unbelievable potential for gold and silver.

"When I started finding out that there was something in there that I didn't know about, he kind of gave me a little bit of a kickstart," Peterson said. "'I need to get healthy enough to walk, and I need to get healthy enough to do this.'"

And he did. Little by little he began to turn his health around and started getting outside again, building his strength so he could work in the mine. Every day he does a little more inside it. Currently, he's putting in power, stringing up lighting, and putting in new ore cart tracks.

"It makes it kind of exciting what we're finding in here," Peterson said. "There are some veins of gold back there that are really three ounces of gold per ton."

Peterson's enthusiasm for the mine and its potential is noticeable in the way he walks with a spring in his step through the train car that acts as a bridge over the stream to the other side. As he walks through the mine, his voice takes on a metallic-like sound that reverberates through the rock tunnels.

"Every day I do something," Peterson said. "We had six feet of ice back in here and to be able to start working we had to come in with a chainsaw."

More Valuable Than Money

Peterson's son, Jesse, grew up around gold mining and decided to go into the IT world and work with computers.

"I come up and help him here and there," Jesse said. "But I grew up here, so whenever you grow up doing something, a lot of times all you want to do is get out."

He said he had fun as a kid on the property but didn't want to do hard physical labor for a living. He admitted he wasn't sure if he'll ever get back into mining for gold like his father.

"Now that I'm a little older and wiser, I guess, I kind of miss it," Jesse said. His oldest son recently spent a summer working for his grandfather and afterward declared his intention to join the Marines, adding that the experience would be easier.

"Eventually, I probably see myself coming back and playing a more active role," Jesse said. But for his part Peterson said he never plans to leave the property he bought from Vic more than forty years ago. He added he's turned down considerable amounts of money over the years to sell his property.

"Let's say I sold the whole thing, and I had a stack of money in my pocket, and I went somewhere—what would I do?" Peterson asked. He looked to his son and repeated the question.

Jesse laughed, shook his head, and said, "I don't know." Peterson said what he has on his property is more valuable than the money someone could offer him for it. He has the mine, he has a placer shaft that he's building next to the stream, and he has the gold-panning business for the tourists.

"If I took $3 million and left here, I still don't know what I'd do. I'm not a big guy to go out and have a lot of fun. This is my fun," Peterson said. "It's never been about getting rich, it's never been about getting another big pocket of gold. [But] there's more out there, and I'll find it."

Peterson sat by a campfire, staying warm in the chilly spring weather. A storm was moving in and the sky has turned an ashen gray, but he paid it no mind.

"I don't know how long I'm going to live, but I figure however long, I'm going to do what I enjoy doing."

CHAPTER 14

WHAT THE NEXT BLAST BRINGS

The stick of dynamite caught above their heads in the telephone wires. The short fuse hissed like an angry snake. "We'd better get out of here," the miner said. With the sort of mutual agreement that comes with the urgency of primed nitroglycerin, they ran.

Ed Lewandowski chuckled at the recollection. He has a sharp memory of the colorful characters that populated his early mining experiences in Colorado.

"He was a little sort of a daredevil," Lewandowski recalled of the miner. "He was showing us around his mine, and some women were with us, and he said, 'Well, you want to know what dynamite sounds like?' So he got a stick and a short length of fuse . . . lit it and said, 'This is perfectly safe, I'm just going to throw this away. It is just going to explode in the air. It won't hurt a thing.'" What could have been the man's famous last words illustrate the exact opposite of Lewandowski's philosophy.

"Not everybody was as safe as me."

And you don't live to a ripe old age as a hard rock miner without being very safe. Lewandowski got into mining just after World War II, in the early '50s. I met the veteran and engineer at a greasy spoon in the Denver suburb of Lakewood. I arrived for our interview a little early, took a seat, and asked the waitress if she was familiar with the person I was planning to meet.

"He's in his mid-nineties; I believe he's a regular here." I didn't want to miss him coming in the door, and at the same time I also didn't want to harass every patron of the business who looked northward of eighty years old. The waitress shook her head, indicating that many of their regular clientele could fit the vague description I gave her. But when he walked in through the front door, somehow I knew I'd spotted him. There was something about the way he carried himself that had the look of an old hard rock miner. I introduced myself, we shook hands, and sat down at the table. Meeting Ed Lewandowski was not unlike meeting mining royalty. No one has been in the business longer. Lewandowski was tunneling under the earth even longer than Al Mosch, with whom he's been friends for more than fifty years.

The retired miner has a great sense of humor, several nuanced layers of expertly defined sarcasm at his disposal, and a memory that could go back decades in an instant, recalling people, facts, and anecdotes. He relayed some of the good counsel he received from the old-time miners, who were on scene as he was originally getting started. One piece of advice he shared with me was to stay away from a certain mine near the town of Georgetown.

"Don't ever go near that mine. It's killed more people there," he recalled. "This was back in the '50s when they were telling me. I started reading some reports and [that mine] killed probably a half a dozen miners back in there because of bad air."

As he told me this, I felt the hairs on the back of my neck rise. Only a few years before while working on an article, I had been in the very mine Lewandowski described. And it was the only time where I felt woozy, my thoughts muddled as I walked around the mine's tunnels with one of the new managers. I remember when I got back to the entrance gulping in fresh air, trying to clear my head.

"You come out all right?" he asked when I shared with him my experience. He then shook his head. "You're lucky you made it out." Indeed—and with that subtle revelation the shockingly gargantuan breakfast burrito I ordered suddenly lost some of its magical luster.

Lewandowski's gold mining mentors were first on the scene in the

1920s, and their mentors worked a shovel and pick in the 1880s. The waitress, it turned out, did know him very well, so much so that he didn't have to order: she simply brought him his breakfast. He ordered a no-frills meal of eggs with a sliced tomato on the side.

"Very good, you did well today," he said, complimenting the waitress.

"Oh, thank you," she replied, then turned toward me. "He is so easy to please; it doesn't matter what he gets."

Lewandowski was born in the early 1920s but doesn't like to share his exact age, happy just to let you know he is now well into his nineties. As he ate his breakfast, slowly and methodically, I asked the oldest living miner that I could find in the state of Colorado about his early days of looking for gold in the Rockies.

"It's fun, it is," he said, then added, "especially when you see the gold coming out." I was soon confident that he had seen more of it than most.

Bringing in a Prisoner

Born and raised in Wisconsin, Lewandowski remembers spending his summers as a kid being loaned out to work farms during the Great Depression.

"I spent a lot of time on farms outside. That's where I got my love of outdoors."

He said when the war broke out he enlisted in the Army Air Corp, the predecessor of the Air Force, because of his long-standing passion for airplanes. He soon became a crew chief and served in a variety of different types of classic World War II–era aircraft such as the AT-11, B-17s, and B-29s.

"That first Christmas was really interesting because we were all wondering who is going to be here next Christmas," Lewandowski said, since the troops were readying for the D-Day invasion. He said a lot of his fellow soldiers operating Waco CG-4A combat glider planes never made it back.

"Well, after [the invasion] they didn't need any more glider pilots, so we're going to concentrate on bombing Germany, so they put us in B-17s. So I was a crew chief on a B-17 for a while," Lewandowski recalled.

"Then when it looked like Germany was on the ropes they said, 'Well, we'd better turn our attention to Japan. So they put us in B-29s, so that's where I ended up as a flight officer on a B-29 when the war ended."

When he left the Air Corp, he went to college and earned a degree in engineering with a minor in geology, and met his wife. The Department of the Interior said they had some openings in Colorado, which he remembered enjoying during a visit to the state while he was in the military. In 1949 the two moved to the Centennial State, and Lewandowski remembered it being not quite as populated as it would later become.

"That's when this area, the Denver metropolitan area, had a quarter million people. It was a delight to get around. And you could go anywhere you wanted in the mountains, anytime. It was just fun."

Lewandowski moved to Colorado to build dams across the West, but during his time off he became interested in climbing, hiking, and history.

"I got to thinking, 'You know, these rocks—I should know more about them.' So I took a couple more courses in geology," Lewandowski said. "Then I got knowledgeable with some miners, and they said, 'You're finding a lot of veins. All this time you're doing it, you ought to start investigating how to do some mining and prospecting.'"

So he joined the Clear Creek County Metal Mining Association in the early '50s. The group met once a month in the basement of the Idaho Springs city hall, which also served as the local jail.

"We'd have a meeting—and every once in a while the police would bring in a prisoner," he said. "Those were the days."

Radioactive

"Actually," Lewandowski said, his fork pausing for a moment over his breakfast, "what started us was the uranium boom." In the 1950s would-be miners and prospectors headed out into the American wilderness in search of uranium for the government as it rolled up its sleeves to deal with a cold war.

"So we all had Geiger counters," Lewandowski said. And after a few false starts, he and some partners came across what he described as the mother lode of uranium near Peaks Grays and Torreys—both hugely popular with visitors looking to tackle a fourteener today in Clear Creek County.

"We chipped off some samples and took it down to the Atomic Energy Commission and at that time it was the only one buying uranium ore, and you had to go through them," Lewandowski said. "So I took some samples to them. They said, 'We don't know where you're getting this, but it is some of the richest uranium ore in the Front Range.'"

He and his partners were encouraged and went back to their mineral claim. He remembers one time on the side of a cliff blasting rock out and watching the boulders flying around him with no place to get to safety.

"I said, 'This is not all that great.'" But they did find additional samples that they took again to the energy commission. He said they were interested in getting a contract and selling what they found to the government. However, supplying the government with uranium wasn't meant to be. They were ultimately told that they could indeed start selling it—once they could provide proof they'd already sold one hundred tons to the government the year before.

"I never could figure that out," Lewandowski said. "They didn't want us little guys; they were looking for big guys. So that kind of scuttled that." He said they went down in elevation slightly and began to find good veins of silver ore. In 1955 he and his partners decided to incorporate. They just needed a good name for their company.

"I said, 'Well, have you ever looked where we've been working? There's [old] broken ax handles, broken pick handles, broken shovel handles. Why don't we call it the Broken Handle?'"

Working at the Department of the Interior building and designing dams during the day, Lewandowski was also working on his Broken Handle mining business on the side and weekends. In time, he retired from the government and focused solely on mining. He said going from building a dam to digging mines isn't that much of a stretch.

"When you're building dams you're primarily interested in a good foundation, which means you're drilling and blasting in the rock. It is just natural." He added with no small degree of pride that during the thirty-five years of dam building and some thirty-five years of working underground, he's never had an accident.

"I got somebody up there looking after me." Lewandowski added he was also a stickler for safety, and every morning before sending in his employees to look for gold they would have a meeting.

"We'd sit around, drink coffee, and eat donuts, having a safety meeting. 'You're going to be drilling, you're going to be mucking—but be sure to do this and this . . .' and we never had a lost time accident. Never."

In the early 1990s, a longtime miner, Charles "Choppo" Fetterhoff, was looking for help in working his gold mine, the Hidee, in Gilpin County.

"He says, 'I think there's a lot of gold around here.'" And it turns out there was.

One Hundred Pounds of Gold

Not as old as some mines in the area, the Hidee began its life in 1896. Lewandowski agreed to help Fetterhoff, and the two began to research the area and the other neighboring mines. Soon he began to see that the area did indeed have a lot of potential. The two continued to look for gold but also decided to add a tour business to the operation.

"We started developing the Hidee Mine then with what little capital I had and started adding to the structures," Lewandowski said. "Then he passed away, and his heirs asked would I mind taking over and running it. Sure."

He said his gold mining partner was long suffering from miner's consumption, another name for silicosis, an affliction that eventually took his life. In time, Lewandowski developed the tourism portion of the mine but continued to drill religiously and blast for gold.

"And we hit the vein. It was beautiful gold." Once again, Lewandowski wanted to go into full mining production but needed several million dollars' worth of investment to make it possible, which he just didn't have. So instead he sold "specimens" of rich gold ore, which he talked about over his breakfast with great reverence.

"I'd get fifty-pound specimens, eighty-pounds, one-hundred-pound specimens of this beautiful gold ore," he said, adding that he ended up donating many of them to local museums. When he turned ninety he realized the time had come to pass the torch to a new generation. Just a few years before, Lewandowski sold the Hidee to a new group, who continue to give tours, and he decided to hang up his pickax and headlamp for good.

"I was still enthusiastic and had the energy [twenty-six years ago]," he said. "The more we drilled, the more we blasted, the better the ore got. That's the exciting part; the exciting part is drilling and blasting and waiting for the fumes to clear and going in and seeing the new bright chunks of gold ore and it was rich." With a shrug he admitted that most years they broke even, but his miners came back year after year to work for him because they were having fun and enjoyed what they did.

"To me, the excitement is, 'What is the next blast going to bring?'" Lewandowski said he knows there are people today with the same hopes and dreams to find gold in Colorado's mountains that he'd had. "I just say to them, 'Well, good luck to you.' I don't try to discourage them." But he hastily added that there are more permits and regulations required than ever before. He is, of course, entirely correct. For better or worse that is what breaks the back of many hopeful mining operations. Over the past several years I'd met and talked with several mining operations that had the gleam of gold fever only to find a sobering cure in the current environment of stringent regulatory requirements. But as to the potential of future fortunes of gold locked away within the mountains, Lewandowski had a quick answer.

"There's no doubt about that."

As our interview concluded, Lewandowski stayed behind, saying he wanted to sit awhile. He was in no hurry. I left the old miner with a handshake and headed back out onto the busy morning streets of the city, while he sat with his memories in the diner's empty booth.

CHAPTER 15

THE LONG TUNNEL

With an eager step, Bob Bowland made his way around the mining, milling, and assaying equipment spread out on the dirt lot below the remains of the Argo Mill and transportation tunnel. The machines are placed in orderly rows, their shadows all leaning sharply to one side in the early February morning light. The area looks like a burial ground for the Industrial Revolution.

Located on the east end of Idaho Springs, the massive red building looming over the city is unmistakable. Twelve stories high, the old mill leans on several huge, yellow waste rock piles overlooking the valley and the town. While it is a stunning relic of the area's mining era, the true wonder, and source of a deadly tragedy, is hidden behind it.

In his seventies, Bowland was boundlessly energetic and pointed out aspects of the historic property. He grew up hearing the colorful stories of the Argo from his relatives, many of whom worked at the site during its prime. Today he's one of its new owners, but its mining days are long gone. The former president of the local historical society was driven in a pickup truck up the side of the yellow waste pile along a narrow, steep, dirt road.

Opening the passenger door, he got out at the top and walked along toward the Argo Mill and under a small entranceway to the buildings beyond. A cool, dry wind blew along the ancient corrugated steel and pushed before it faint ghosts of dust. Past the empty buildings and forgotten milling works, there's a giant sealed hole in the side of a mountain. It leads to a tunnel more than two thousand feet below the surface, which stretches for more than four miles to Central City under some of the richest mining in the area.

An incredible feat of engineering, the Argo Tunnel represents the peak of the mining district's productivity and its ultimate and fatal decline. Where there is stillness now, there was once a flurry of activity and noise. Bowland walked along where trains once ran, where hundreds of miners were transported every day into the massive subterranean tunnel and where rich gold ore was once removed by the tons.

The Mighty Argo

As mining for gold in the streams of Colorado moved into the mountains and the underlying rock, miners started to experience new and challenging difficulties. The deeper a mine went, the harder it became to remove the rock, but even more troubling was the water that began filling the tunnels. Mining companies would often employ early versions of water pumps to clear the tunnels, but as the mine went deeper, so too did the water.

The future of mining looked grim. Some of the richest areas of all including Quartz Hill, once called "the richest square mile in the world," needed a solution that would continue to let the mines operate.

Entrepreneur Samuel Newhouse was presented with a solution to the water issue, and once he secured the funding, work began in September 1893 to build a massive drainage tunnel.

A writer for the *Silver Cliff Rustler* reported that year, "The Argo Tunnel is a gigantio [sic] enterprise owned and operated by Samuel Newhouse and some European capitalists. It has for its objective the discovering of the riches of the famous Seaton Mountain, Russell Gulch, and Quartz Hill."

The newspaper estimated that about $100 million was removed from the area from 1861 to 1893. "A vast army of men will be employed, and the output from the mountain will no longer remain at $3 million [a year], but will increase many fold."

Work began on the Newhouse Tunnel, later renamed the Argo Tunnel in 1893. The area's iconic gold mill was added in 1913.
(Courtesy of the Historical Society of Idaho Springs)

For a fee, mines could continue to remain sustainable by connecting to the tunnel and draining their water. In addition, miners could dump their ore into the tunnel, then load it onto trains for transport to the area's mill for processing. The newspaper continued, "It will afford not only a cheaper method but the cheapest method known in mining engineering for working the famous mines of Gilpin and Clear Creek Counties . . . [which], originally gave to Colorado its wealth and reputation."

The workers needed to be able to breathe and operate their machinery, and so the first thing built was a compressor plant, which could force air for four miles. Work started on the tunnel with hand drills, then moved on to the dangerous pneumatic air drills for the tunneling work.

Years later mechanized drills finished the tunnel. Power stations were built to get electricity to the mine for its lights and electric locomotives. Because of this, Idaho Springs was one of the first in

Colorado to get electricity. Men who drank too much or didn't have the backbreaking stamina necessary were immediately fired from the project. The tunnel was going to be a massive undertaking, and fortunes were at risk. Miner Merle Sowell, who worked in the tunnel years later and knew men who had worked there before him, wrote in his book *Historical Highlights of Idaho Springs* that there were plenty of ways a man could be hurt during the building of the tunnel. After the tunnel was dynamited, men would go in with scaling bars and shovels to clean up the mess. In the rush to clean up, workers were often badly cut or suffered broken hands and feet. Falling rock could instantly kill a worker or leave him terribly injured for life.

Early estimates said that drilling about a mile a year, the tunnel would be completed in as little as four years but that wasn't meant to be. As the years went on, the project grew ever-more costly and Newhouse's substantial resources were taxed. Then as the project bravely entered the new century, it ground to a full stop. The tunnel was three miles in and on the edge of its most valuable mining areas. Boston banker Frank Schirmer became interested in the project and took over the work from Newhouse. Schirmer put up the money and pushed the tunnel through to the end. The tunnel was completed in 1910, and a mill was added in 1913. Originally named for its benefactor, the "Newhouse Tunnel" was now renamed the "Argo Tunnel." Schirmer then put the project into the care of his younger brother and mining engineer, Rens Schirmer.

The tunnel was built on a 0.5 percent grade, which helped to drain water from the many mines running into the tunnel. The water passed through a ditch built in its floor, which then ran outside. The tunnel also had trains shuttling in people and bringing out ore on two tracks.

A five-hundred-volt direct current trolley system was used and two seven-ton Westinghouse locomotives were employed to haul out the material, pulling as many as 45 to 102 cars behind them. Rens Schirmer said the tunnel was like a train railroad.

It's hard to imagine today among the dust and rusted buildings that there were once sounds of trains and miners loading up to go to work. Like a trolley it took them into the tunnel with the jerking efficiency and noise of iron wheels on narrow-gauge railroad tracks. The tunnel was much colder on the inside and the wind blew in one direction or the other depending on the time of day. A miner might sit on the train with his equipment watching the tunnel's entrance sink into the earth as the track continued up on its slight incline. The other train would pass by going in the other direction carrying ore to the mill or to another train that would bring it down to Denver. It could be several miles into the tunnel before the breaks would shriek to a stop and he had to jump off, making his way to the mine employing him.

Hundreds of men rode the specialized train into work, but it didn't just transport miners. Beatrice Rule, a young teacher at her first job, also rode the train into the great tunnel with miners to get to her isolated little schoolhouse. After about a mile she was carried by a hoist up the shaft of the Gem Mine to the Gilson Gulch School. At one point the tunnel served close to one hundred mines from Clear Creek to Gilpin County. Men worked to expand tunnels and remove all the valuable minerals they could find. It wasn't all gold and some of the ore, such as uranium, was just not recognizable as valuable to the miners at the time.

In some shafts, such as the Pozo Gilpin, the radioactivity emitted from the uranium made the rock so hot that the men couldn't work as deep as two hundred feet below the surface of the Argo Tunnel. It is said that Madame Curie received her first pure sample of the radioactive rock from the Blucher Mine on Quartz Hill.

The Argo Tunnel was extremely lucrative and extended the life of the mines, in some cases, by as many as thirty years but nothing lasts forever. After World War I, disputes among the tunnel's owners stuttered business. The tunnel was cleaned up and readied again during the late '20s and bought by mine owner George Collins, who had big hopes for the mighty Argo. By the 1930s, all the other mines had finally come to a close. Collins had hoped to use the tunnel to come across the Kansas shaft, which was known for its gold but left untouched for decades. The mine had stopped pumping out water under the assumption of a connection with the Argo Tunnel.

Collins knew the mine was left to flood and planned to drain it into the tunnel. It was muddy, dark, and dangerous work, and with a skeleton crew, they plunged through the rock looking for the mine's shaft. Sowell said that working in the Argo and searching for the Kansas shaft was like working in a freezing shower with all the water pouring in on the workers. Men who could find drier jobs quit as soon as they could. A rich vein was discovered, validating the hunt for the Kansas shaft. Because of the water pressure, miners often had to dodge high-powered blasts coming through the rock at what was estimated to be five hundred pounds per square inch. The miners were concerned about what would happen if they came across a shaft full of water but continued their work. Luckily, Sowell left for a drier and safer job, while the work continued for two more years.

January 19, 1943

Winter in the Rockies brought with it the teeth-chattering cold and undoubtedly made mining in the mud and snow an unpleasant business. Under Collins's leadership, the men, now only four of them, continued to drive a tunnel, chasing the gold vein ever farther into the mountain. As they worked their way up a stope, the water pressure, squeezing through the rocks, began to increase. The miners used dynamite to continue their progress. After a time the pressure didn't increase. The miners believed they could use a shorter dynamite fuse. It was a fatal decision.

Bill Bennett was operating the electric train and was headed out of the tunnel when the lights flickered and went dark. With no power, the train came to a stop. He was about three hundred yards from the entrance. With the train now silent Bennett could hear a distant roaring. It got louder and louder until the rumble was deafening. Bennett correctly assumed what had happened, climbed down from the train, and ran for his life. Before long the water reached him, and he continued running as hard as he could. The water crept behind him first to his ankles, then his knees, and by the time he got to the entrance, he was alternately swimming and running through a waist-deep torrent. He got outside and sounded the alarm. Soon the water completely filled the tunnel from top to bottom and ran nearly seven hours straight.

The four men inside were dead.

Collins had visited the mine earlier that day and was on his way to Denver on a bus when he noticed the river turned color and began to suddenly rise. It is said that he stopped the bus and hitchhiked back to the Argo Tunnel. He saw the devastation, the hard looks and harsh words of those already there, and nearly collapsed from the shock. A dispatch from Idaho Springs, which ran in the *Steamboat Pilot* four days after the accident, reported that the last man was finally recovered near the other end of the tunnel, under five feet of muck. The Gilpin County coroner determined the men had died due to negligence.

"And so that closed the tunnel down," said Bob Bowland. "It tore out all of the ladders, the track, the wire. There are probably a couple hundred ore cars that are tumbled around and rusting away and timbers and track and wire and machinery—it's a disaster. A colossal disaster."

With the severe damage to the tunnel and impending lawsuits, the state closed the mighty Argo for good.

A Deadly Legacy

And so the sound of men and their machines went from a rattling echo to a fading memory. The tunnel, mill, and the many mines have now been long silent.

But they did not sit still. The tunnel, as it was engineered to do so long ago, continued to drain water. Day and night the many mines connected to the Argo filled the giant tunnel, draining out onto the yellow waste piles and into the stream below like water through missing teeth. Several attempts were made to clean and open the tunnel, but the damage was too severe and all efforts were ultimately doomed to failure. At some point a massive metal door was put in place to bar the tunnel's entrance.

But water continued to drain from beneath the door into Clear Creek. Over the years something sinister happened to the water. It became acidic. The process went like this: When the tunnel and conjoining mines were built, pyrite, an iron sulfide, also known as fool's gold, was exposed to oxygen. Pyrite in the presence of both oxygen and water essentially forms sulfuric acid. That acid lowered the pH of the water and caused the dissolution of the other heavy metals in the surrounding rock. Minerals such as iron, manganese, zinc, aluminum, and copper began entering the stream—most in very high concentrations. The water also saw elevated levels of cadmium, lead, and arsenic. The mineralization had always been there, but the Argo allowed oxygen to infiltrate the tunnel, causing acid mine drainage to form.

In a single minute as many as five hundred gallons of this acidic water, which had the same pH level as vinegar or lemon juice, was entering the stream. The acidic environment killed off many of the fish and other aquatic life, polluting the stream for thirty-five miles, until it merged with the South Platte River. The stream running past the Argo Tunnel is also the main source of drinking water for more than a quarter million residents living in the Denver area. The contamination coming from the Argo Tunnel was estimated to be anywhere from one-third to one-half of the total metal load carried in Clear Creek.

In 1983 the area was listed as a state superfund site. Nearly eight years later an eruption of contaminated water from the tunnel known as a "blowout" occurred, which turned the stream a sickly shade of yellow for fifteen miles. Several downstream municipalities had to close off their water plants from the contamination. Something had to be done.

Mary Boardman is the project manager of the superfund site and water treatment facility created to mitigate the Argo's acid drainage. She said early on the Environmental Protection Agency (EPA) tried to see if there was another way to deal with the contaminated water, maybe find a way to seal it off or divert the clean water around the area.

"In the late 1980s the EPA had some of their contractors go into the Argo Tunnel to see if they could do any source control . . . they didn't make it very far in," Boardman said. "They only made it a couple of thousand feet back into the tunnel and just decided that it was not possible to rehabilitate it or do any type of source control."

The treatment facility was constructed in the '90s. From outward appearances it is almost as large as the Argo's mill and has a similarly designed exterior, but they are worlds apart. Inside, a host of tanks and processes are in place to treat the contaminated water running out from the mouth of the tunnel. The outflow of the tunnel is monitored twenty-four hours a day. Boardman said the water's treatment is relatively easy.

"It really is a simple process," Boardman explained as she navigated the facility's raised metal catwalks suspended high above several giant treatment tanks. "Basically the water is acidic so the metals are dissolved, we raise the pH up to precipitate the metals as metal hydroxides, and then we remove them from the water. So real basic chemistry."

With a handful of contractors, the facility works day and night to filter the water coming out of the tunnel, removing nine hundred pounds of contaminated material a day before letting the water return to the stream. The hazardous material is then transported to a landfill. Approximately 350 gallons a minute is treated on average, but the building can handle as much as 700 gallons a minute.

If another blowout should occur, the water plant's operators can remotely seal shut a newly installed bulkhead so as not to overload the current treatment system. The massive bulkhead was installed ninety feet from the entrance of the tunnel to hold back the contaminated water. A small rounded door in the center of the bulkhead allows a man to crawl over to the other side in an emergency. With a black marker someone scrawled on the door, "Do not open." A pipe runs from the bulkhead carrying the contaminated water to the treatment facility below. Several other local mines and tunnels also have their acid drainage transported through pipes to the water plant. The EPA originally ran the site, but the state soon took it and its funding over. Their serious, and rather daunting, mission is to operate the facility forever.

"On sites where we do long-term water treatment, typically, the state will take over as the lead agency role because we're going to be funding it in perpetuity," Boardman said. "We don't have any other options—we can't just plug it because the water will go someplace else."

Colorado owns and operates several similar facilities across the state, as do other public and private organizations, as part of the cost of the state's prolific mining history.

A Magical Ship

In the 1970s the mill and property was purchased and turned into a popular tourist destination that, for a time, even had pretend shootouts. Nearly thirty thousand people visit the Argo Mill and Tunnel every year.

Inside the mill itself, lit by the dim amber light filtering in through the ancient windows, Bob Bowland moved among the remaining relics. The equipment inside the mill is remarkably intact, but much of it is missing.

"In 1943 when the tunnel flooded and killed the four miners, it all

Operations stopped at the Argo Mill and Tunnel in 1943 after a deadly accident claimed the lives of four men. *(Photo by Chancey Bush)*

came to an end," Bowland said. "Remember World War II was still going on—and a lot of the missing equipment went to the war effort. I wouldn't be surprised if some of it was lying on the bottom of the Pacific Ocean right now."

Bowland is a history buff, and walking through the iconic property, which he now owns, is clearly a huge thrill. In 2016 Bowland and several business partners purchased the property in hopes of developing hotels, houses, and restaurants on the yellow waste rock. He also plans to keep the mill intact for continued historic tours.

While the inspiration behind changing the name all those years ago from the Newhouse Tunnel to the Argo Tunnel remains a mystery, its newest owner has a pretty good idea what its new name meant.

The mill's job was to remove the gold from the ore rock brought to it. It did so by crushing it under giant stamps and using everything from acid, mercury, and even soap to remove the gold from the rock dust.

"Also, lanolin has an affinity for it," Bowland said. Lanolin is the wax secreted by sheep and other animals with wool. "That's where Jason and the Golden Fleece story comes from. The earliest miners back in the

Bronze Age . . . used sheepskin to pour the ore through. The gold would have an affinity for the lanolin and they would either wash out the sheep's skin or burn it or dry it in the sun—so they could get the gold dust out of it."

Hence the golden fleece, and like the magical ship sailed by Jason and the Argonauts, Bowland is hoping the Argo will once again seek out and bring prosperity to the small mining community trying to survive in the shadow of the gold rush.

CHAPTER 16

A FAMILY LEGACY

Nearly three miles due north of the Argo Tunnel, over rough terrain, cratered with the remains of old mining operations sits the Chase Mine. Located in Gilpin County and shoulder to shoulder with some of the most famous mines in the state's gold rush history, the Chase Mine lies in a cone of depression still being created by the drainage tunnel. The Argo, still doing its job after all these years, removes water from the mine's 130-year-old tunnels all the way to 600 feet below the surface.

The Chase Mine is currently one of three hundred mines listed in the area, but between Gilpin and Clear Creek Counties, it has been estimated there could be several thousand. For a portion of its early history, the Chase Mine was one of the top gold producers in the state.

The mine's other major historic distinction is a death that occurred in 1896 when Austrian miner Tom Zadra came across "bad air." Some mines, especially because they're often not well ventilated, can have a deadly buildup of gases, including carbon dioxide and methane. These displace oxygen in a mine's tunnel. People experiencing this phenomenon find they can't think clearly, and their body stops reacting the way it should, their muscles no longer responding. Unconsciousness is quickly followed by death.

An article in the local paper reported Zadra, a seasoned veteran of the mine, was giving a tour of the mine's operation to another man after it had been inactive for two years. At three hundred feet below the surface, both of the miners' candles went out at once, leaving them in complete darkness. With no oxygen, the flames couldn't continue. Zadra warned the other miner that there was a history of bad air in the Chase. The man then heard Zadra fall away into the darkness. He ran for help, making his way through the dark mine. When he returned, Zadra was already dead.

But over the course of the mine's long history, it continued to produce, opening and closing again—and when open, providing modest fortunes for those who worked in its depths. Then about a hundred years ago the mine was purchased by Jacob Smith's family and has remained with them ever since.

Ragged Ass Miners

On a warm Monday afternoon in Denver a group of men and women with cafeteria trays, piled high with food, filed into a reserved room. Located in the rear of a giant all-you-can-eat restaurant buffet, members of the local chapter of the International Order of Ragged Ass Miners were preparing to start their meeting. The organization is the oldest, still-active group of its type in the state. Established in 1891, the group brings together geologists, miners, and rockhounds once a week to discuss the industry and its history, as well as to provide educational lectures and networking opportunities, all fueled with an endless supply of barbecue riblets, Texas toast, and tater tots.

Most of the seats were taken in the room before twenty-four-year-old Jacob Smith arrived. Dressed in a large olive-green military-style jacket, swamp hat, and glasses, Smith looked contemplative and was every bit as educated as his enrollment in the Colorado School of Mines, one of the top institutions in the world for mineral and mining engineering, would suggest. Smith was welcomed by the group and made an official member of the Ragged Ass Miners after he and several others, including myself, were voted in.

For those who came over to his table to visit, he showed off several specimens of turquoise and phenakite crystals that he'd recently found while digging in the hills. Smith is a geological specimen enthusiast, and he's in good company. Other members carried in their pockets various mineralogical samples that had caught their attention. One man cradled a large chunk of iridescent rock with obvious pride.

Smith was only months away from earning his graduate degree and beginning his future—a future he believes will include his family's gold mine. "The Chase," as he calls it, has served his family since they acquired it more than a hundred years ago.

His family had moved to Central City in the 1860s to follow family patriarch James Willis, who worked as a gold miner in the Fourth of July and America Mines.

His daughter Mary Collins, born in 1857, described the trip her family took to be with their father during the early years of the railroad in Colorado. She recounted a long, difficult journey, and near the end she recalled how the train derailed and rolled down an embankment with everyone inside. None of her siblings or relatives were seriously hurt in the accident, but two men had died. Her brother and Smith's great-great-grandfather, H. C. Willis, later bought the Chase in 1911. It closed for a time in the 1920s after one of his great-grandfathers, then running the operation, died from a staph infection.

"So once he passed away, he was really the driving force behind the mine, and they shut down," Smith said. During the Great Depression, his other great-grandfather processed a lot of the mine's dump material with a shovel over the course of several years to make a living.

"That was his work for a couple of years during the Depression when there was nothing else to do," Smith said. "He found gold and it was sustainable at the time. He was the last one to mine it."

Then in the last several years, Smith and his family have worked to get the mine back open again. They cleaned it up and opened its tunnels. Smith has been inside, and he has seen gold.

"There's a couple of million dollars' worth of gold in there."

A Passion

Smith grew up going on rockhounding and arrowhead-collecting trips. But during his freshman year of college the subject really set fire to his imagination, and he decided to go for an undergraduate degree in geology.

One teacher's recommendation led to another and finally to a look at a museum where Smith became infatuated with minerals and fossils. He decided to attend the Colorado School of Mines and earn his graduate degree in Mineral Exploration and Mines.

He plans to have the family's gold mine open again. Smith has poured over all the old records and reports of the mine, and in addition to seeing the gold in person, is sure that there's plenty left. Not only does the mine have gold ore rich in content, he believes it also has both rare earth minerals and the possibility for the ever-elusive wire gold.

"As the geologist on-site I want to make sure that stuff doesn't get smashed. A lot of it belongs in museums," Smith said, adding gold specimens like that are often worth two or three times their weight. "I just want to be there when the mine is open to make sure that those don't get crushed, because if you look at a hundred years ago almost all of it got crushed and maybe there's a handful of specimens that didn't."

The Future

Gold's popularity in the annals of Colorado's history is self-evident with the countless number of yellow mine waste piles streaking down the hillsides of the state's well-known mining areas. But its future, according to Dr. Richard Goldfarb, is less than certain.

"What we actually mine changes with the price of gold," Goldfarb said. When the price of gold is very high, he said it's worth the effort to remove even the faint traces of the mineral from rock surrounding what may have once been a gold deposit. "When gold goes to $1,800 an ounce, it becomes ore. When gold goes back to $1,200 an ounce, it is just altered rock," Goldfarb explained.

"The potential reflects the economics. You see a lot more people prospecting Colorado when the price of gold is very high." He added the potential for alluvial or placer gold increases every time there's a major storm that reworks the soils. However, Goldfarb believes the days of major Colorado gold mining will live on only in its past.

"The potential just isn't there because of the types of gold and also the cost of gold. To make a gold deposit economic depends on the environmental laws and the cost of labor." He explained that gold mining companies are more likely to explore a country like Myanmar over a state like Colorado.

"Because it might cost a $1,000 an ounce to mine here where [elsewhere] it's a $100 an ounce," Goldfarb said. "The big companies don't want to take a chance on something that is going to cost them $1,000 to $1,200 an ounce to mine." If gold is worth $1,800 an ounce, there are places where it can be mined less expensively so that the company is assured a larger profit.

"They don't even think about the geology. They think about the countries and throw out the countries that aren't going to work. Many companies will throw out the US for various headaches, especially the expense of the environmental issues." He said the demand for gold continues to be enormous, and in many parts of the world gold is considered a better bet than currency.

"Gold is a mineral that is not extremely abundant, so it makes it very valuable, and historically it was used instead of hard currency. And the Asian countries, still, they don't trust hard currency, so gold is in high demand. As countries like India and China get more wealth, the demand for gold has gone way up."

Goldfarb said that depending on the estimates, at its current demand of 2,500 tons a year, only eighteen to thirty years of gold resources remain across the globe. With the price of gold down to $1,223 an ounce, he added that there are also no new gold discoveries and companies are laying off their exploration staff and just working old properties.

"So eventually we're going to have to make new discoveries, or we have a concern here," Goldfarb said. "Demand is increasing, we're not discovering more, and production is the same—something has got to give."

Getting the Gold

Smith, who once took classes under Goldfarb, agrees with his former professor's assessment of the future of gold mining in Colorado, but that doesn't dissuade him. He believes there is still a bright future for small-scale gold mining operations like his.

Smith added many of the gold veins in Gilpin County and other places in the state were both too small and not conducive to the current trend of open-pit gold mining.

"The interesting thing about gold in Colorado is . . . these types of ore deposits and mineral deposits are overlooked by major companies—but for individuals they still make sense."

He said a small operation could remove a few tons a day from a mine and process the old dumps and leftovers in the mine and still make a very nice profit.

"If you have a place to process that, it makes a lot of sense," Smith said. "And the mines are already in families—there are people who would be able and are ready to work those."

Smith is 100 percent positive that his family's mine will be economically feasible once they can reopen it.

"Most of the people that are in the industry are looking for a gold deposit. They're trying to find the gold. I am in kind of a unique position because I found the gold." But he now has to figure out how to scale up the production to extract the gold from the rock.

And that's the trick. Before Smith can remove gold from the Chase, the mine needs a place to take its gold ore for processing. Most gold taken from a mine needs to be removed from the rock that it's in. Rarely does a miner find pure gold deposits like the men in Al Mosch's story about the Lamartine. Colorado once had many places a miner could take his gold ore for the milling process, but today there are virtually none for small-scale mining.

"It's one thing to get the mine opened and permitted and ready to ship ore—but if you don't have a place to ship that ore to, then you're dead in the water," Smith said. "Even if you have the highest-grade ore in the world, if you don't have a place to send it . . . you won't be working."

Smith and his family have been looking for a viable gold mill for years, and he said there are several on the verge of opening but none have yet done so.

"The mine is pretty much ready to go. There're a few thousand dollars of permitting that needs to be done, but mostly it is [lacking] a mill," Smith explained. "I would go out there and dig the material out with a shovel myself—moving the rock is not a problem—it is just the processing of it."

Smith also believes that a gold mill operation in the state could help turn a lot of gold-rich mine dumps into profitable operations. Because gold-removing techniques were relatively primitive a hundred years ago, many of these piles still have large quantities of gold waiting to be discovered.

"Even if this is something fifty to one hundred years in the future, eventually we're going to run out of super-giant gold deposits and have to think small," Smith said. "I don't know if it is ten years in the future, or a hundred, but mining will come back to this part of the world."

CHAPTER 17

PHANTOMS

Between Central City and Blackhawk, two of Colorado's most famous gold mining districts, sits what is left of the Lily Belle. The mine is just one of what is estimated to be some six thousand mines located between Clear Creek and Gilpin Counties. It doesn't look like much now. Its two entrances, not far from Gregory Street, are collapsed, years ago mitigated by the state's Division of Reclamation, Mining and Safety's Inactive Mine Program. Once hidden by a long-ago demolished building, the mine had an entrance and exit that met in a U-shape. Local legend has it that inside those tunnels the Lily Belle also had a ghost.

The story takes place during, or not too long after, the gold rush. A man had spent the last of his money on a mining claim. Mining and gambling are not all that different, each with followers hoping to hit the jackpot and both sharing their proclivity for disastrous ruin. The man's wife is said to have gotten into an argument with him about his choice. Supposedly he then chopped her up with a cleaver and hid her in the mine to keep her body cold until he could dispose of it. Before the mine was closed, people who went in said they could feel the temperature suddenly drop or, in some cases, even see the ghostly specter of a woman staring at them from the mine's deep shadows.

Old mines, like the ghosts of local legends, have the nagging tendency to come back to visit the world of the living. Not far from the Lily Belle, just a little farther up Gregory Street, an ominous sinkhole appeared in an empty dirt parking lot. Like a yawning mouth, the hole continued to widen, deepening and growing larger. In the summer of 2016, Deb Zack, project manager for Clear Creek and Gilpin County's Inactive Mine Program, was called in to find what was causing the hole and to put it to rest for good.

A Dangerous Surprise

Two weeks before I met with Deb Zack, she and some of her colleagues stared down into that pit as an excavator dug deep trying to find the source of the problem. Already fifty feet deep, Zack needed to find what was causing the problematic hole to form. She said the old mining district is riddled with an uncountable number of old tunnels, shafts, adits, and winzes (tunnels between mine levels) that crisscross every conceivable portion of the area. Another shovelful of rock and dirt uncovered a new hole at the bottom of the pit. Ancient timbers were revealed, and Zack peered down into a stope, where miners had once created an underground void from which they removed the gold ore, going down an additional thirty-five feet. She needed a closer look to better understand where the bottom truly was. With a hard hat and climbing gear, Zack climbed onto the excavator's shovel and let the driver lower her into the hole.

"Which is pretty safe. If you have a good operator that you trust," Zack said. Only enough light reached into the new hole to reveal wooden beams; the rest was lost in darkness. "I was concerned, I wanted to know where the bottom was. It had to be going somewhere," Zack said.

Zack's job is to close mines and associated holes in Colorado's two oldest mining districts. She needed to know where the bottom was so she didn't direct workers to place rock and cement on top of something that would later continue to sink. Suspended above the bottom with a flashlight she peered into the void, looking into an area that has never seen the light of day and hasn't been seen by human eyes in more than a hundred years.

"That was a fun day," Zack said. She added underground voids created by mining could be a dangerous surprise. "You never know what you're walking on top of, especially here," Zack said, waving to the mountains surrounding us. We sat in a parked car just a few feet away from where that hole was recently filled. "Look around; we're in a valley with who knows what kind of crisscrossing claims—I mean this whole area is Swiss cheese. It just is."

Zack recalled the story of two guys recently walking along a mining area a couple hours south of Gilpin County; one man stopped to look back at the other, and he had simply disappeared. He'd fallen into a void.

For a state created because of its mining wealth, Colorado has a pretty fair share of mining-related accidents and deaths. The state's Department of Transportation sees its fair share of headaches as well. Interstate 70, which reaches up from the Front Range into the mountains, and to the famous ski resorts beyond, is closed every several years when a mining-related hole appears in one of its lanes. The state's mining history is often buried and forgotten, but it doesn't remain that way for long. That's where Zack and her colleagues with the Division of Reclamation, Mining and Safety's Inactive Mine Program come in. For the past eight years, Zack has been assigned to Clear Creek and Gilpin Counties. Over that time she has closed some three hundred mines and mining-related holes.

A "rural project manager," Zack said her area of expertise is hole closure, safeguarding the public, and working with Colorado's Parks and Wildlife to ensure that bats colonizing abandoned mines can freely enter and leave. Her organization's history really begins in the early '80s when a fleet of people were sent out to see how many abandoned mines existed in the state. They then created an inventory.

"It's fairly scientific because they would go to documented mining districts where there was known activity and just canvass, and hike it," Zack said.

The problem, of course, is that finding every mine is nearly impossible given the near hysterical rush that followed the discovery of gold. Additionally, just because a mine might have been built with one entrance doesn't mean time, weather, and erosion don't add others. Zack said she and her colleagues, who work in other regions of the state, will work on mines found during that initial inventory. They will be called to close a mine-related issue like the one in the parking lot near Gregory Street, or, as is often the case, will get a phone call from a homeowner concerned about something they discovered on their property.

"Those are always fun. 'Hey, there's a big hole in my backyard now,' or 'Hey, I live in Texas, and I bought this huge piece of property, and there's a giant hole on it,'" Zack said. When she travels to investigate a mine to close, she scouts out the area and looks to see if there are others that, with a property owner's permission, she can also close. "Because I don't want to close just one hole—I want to do twenty if I'm in the area," Zack explained.

Like a Bottle of Wine

I met Zack outside a convenience store with the proud logo of a Sasquatch. The store was located at the start of Central City Parkway, just east of Idaho Springs, a road created to better ferry gamblers from Interstate 70 into the casino towns of Blackhawk and Central City.

Although she once drove a jeep, Zack had to give it up for a state-assigned four-wheel-drive Toyota hybrid. She made what she called a tough choice to save the wear and tear on her neck. Many of the roads she drives were planned and built long before the advent of the automobile. And while the jeep rarely got stuck on her travels, it did put her into a state of constant whiplash.

As we drove along the parkway on our way to a ghost town, Zack pointed out the window at the gated-shut Young Ranch Road. She said there's a local story from about 1900 of a man who used to live in that area and come into town on occasion with a little bag of gold—just enough to pay for supplies.

"Just a real low-key guy," Zack said. "He'd buy some supplies and head on back up. He never shot his mouth off about it or anything, but he picked up some attention from some less reputable folks."

The story goes he was followed back to his cabin one day and was held hostage and interrogated as to where he found his gold. But he wouldn't tell his captors. Time passed, and no one knew what had happened to the man, and after a few months they decided to send some folks up to check on him.

"They sent a search party up to look for him, and they found him dead, still tied up." Which goes to show you that no matter how careful miners are, there is ever-present danger associated with finding gold.

Wherever you turn while traveling through Colorado's mining district, there's a tale or evidence of a fortune lost. At the bottom of the parkway, we turned and drove through the ghost town of Nevadaville. Not much there today, it's one of the many towns that grew from nothing overnight, seeded by the desperate hunt for gold. A few random buildings stand here and there, but Nevadaville once had many stores, hotels, and hundreds of residents. An old photo shows a snow-patched and all-but-treeless landscape populated with homes and residents traveling by horse and buggy. The town once had many fraternal organizations and was well known for its formal dances, music, and masquerade balls.

We drove down an empty road and turned up a mountain to visit the Hidden Treasure Mine, long ago a major employer for the town. We came to a closed and locked gate, so Zack and I got out and hiked up to the mine. At the top of a hill and almost entirely covered in the shadow of overgrown trees lay what is left of the famous mine. Piles of broken wood represented buildings and others made of metal still stood but were rusted almost black.

"There's a seventy-five-foot hole right here," Zack said, pointing at our feet. "Right where we're standing."

In point of fact, the hole no longer existed. At least not one that you can see. In a spot now covered by pine needles, the only trace of the once-giant void was a vertical pipe coming out of the ground.

Zack said years ago people would hike up into the area, climb around the old buildings, and come across a pit that fell away nearly eighty feet. She was called in to mitigate the giant and dangerous hole, which she did with lots of expansive polyurethane foam, which essentially filled it like concrete.

"You can just jump on that stuff, it is great," Zack said, adding it acts like a massive wine bottle cork. The pipe is used to help balance the pressure above and below ground.

Inherently Dangerous

In her kit, she had climbing gear, a hard hat, and a meter that can read for any poisonous gas. She joked that she also brings her "canary dog" sometimes on her trips. Mines are inherently dangerous, and mines abandoned for a hundred years or more pose the greatest risk. It's not just the danger of cave-ins and falling rock but also mountain lions, snakes, bears, and sometimes people of questionable intent.

"The big issue, especially out here, is meth labs," Zack said. "The adits, specifically the horizontal ones, are perfect for a meth lab."

She recalled the story of one of her coworkers who was assigned to neighboring Boulder County. Because their job takes them to mines that aren't necessarily clearly defined on maps and can also have multiple owners, the risk of trespassing happens quite often. However, in one case she said a man jumped onto the hood of her friend's car with a firearm, threatening the kind of extreme violence that can only come from the barrel of a shotgun.

"Yes, it is inherently dangerous," Zack said. "I'm out in the middle of nowhere with no one with me and with nothing to defend myself."

Zack added that because she's just five feet tall and a woman, she often got a pass from people who don't see her as a threat if she accidentally trespassed on their property.

"I trespass like a lot. A lot. Probably more than I should. But am I going to call like fifty different people to say, 'Hey, I'm going to be on your property for like five minutes?' No," Zack said. "But when I do get caught it's nice to go, 'Hey, how's it going? I'm sorry.' You got to play to your strengths, man."

While unexpected animals and people found in and near remote mines do prove a danger, Zack said the most hazardous part of her job is probably the driving portion. None of the locations she drives to is easily accessible. While riding with her to another remote spot, I felt my right foot incessantly pressing on an imaginary brake pedal as we went down a road not meant for anything other than a burro. Trees and rocks scraped at the sides and undercarriage of the Toyota as Zack masterfully navigated a road that looked more like the jagged teeth of a great white shark. Many of these roads don't even have names, while others bear unsettling monikers like "Oh My Gawd Road."

Once she was stuck on the north-facing side of Saxon Mountain above the town of Georgetown. Her car was high-centered on a rock, and she tried to safely chisel her way free without any luck. Hours had passed before some ATV recreationists came across her. The three of them liberated the vehicle from the rock. She said the men were friendly and not creepy.

"I was very lucky I wasn't on the serial killer short list that time," Zack joked.

But she's not kidding when she said the roads she needs to take to get to the mines are dangerous. One time the back end of her vehicle slid out over an abyss. Using the jeep's winch and a nearby tree, she was able to pull it back onto the road. She credits her dad for imparting an appreciation for tools and vehicles.

Inside abandoned mines, she said she uses her slight stature to get around in places that a larger person would find difficult. She also carefully reads the rock signs inside the mine, looking for fallen rock, to determine if it's safe to move around it. Other hazards they'll run into besides cave-ins include coming across items such as old uranium cake. She said the way to avoid most danger is just to use a little bit of common sense. Sometimes in the mines, she'll find old equipment such as candle-powered headlamps. But most of the artifacts are located deeper in the mines, and she doesn't go that far in unless someone else is around.

"When we come out to do the mine closure, and there's more than just me out there, and somebody who would actually know when I've gone missing, then I'll go back in and kind of assess it a little bit more, just for posterity. Take some pictures, take some notes," Zack explained. "A lot of times it is almost a shame because these things are just too dangerous to go back in. A lot of the history that may still be way back in there is just buried."

Recently the FBI wanted to get information from her on a closed mine in her territory. Apparently, this isn't unusual. Hiding in mines or stashing illegal materials in them does make a certain amount of sense. That is until you look inside and realize there are safer and cleaner places to be on the lam. In this most recent case, the FBI was looking for a man and had reason to suspect he was hiding in the Bangor Mine. Zack gave them permission to cut the lock holding the metal gate shut so they could take a look around inside. How it turned out or if they found their man, Zack could never get an answer. She decided to drop by the Bangor and replace the lock the FBI had cut.

With its entrance just below ground level, there was still snow in spots, even in the summer. Zack walked up to the metal gate and using her smartphone flashlight peered around inside. The tunnel was well constructed but certainly didn't look like a comfortable place to vacation, even if you were running from the cops. She noticed light sticks on the ground inside the mine and then saw that the FBI didn't cut the lock but rather the portion of the door needed for the lock.

"Come on now, have a brain," she scolded.

Best Job in the World

Down another road that would function better if used as, let's say, a saw, Zack stopped the vehicle. She told me that she couldn't see the road ahead and wanted to make sure there was some, such as it was, where we planned to drive. After a moment, she got back in, and we confidently drove forward and down onto another portion of the dirt track. The ups and downs of the road were like a roller coaster but one that was always threatening to pull out the guts of the vehicle.

"Yeah, yeah, yeah—I know you don't like that," she said to her vehicle as her undercarriage scraped over a rock. The road we were on isn't meant for cars or any transportation devised in the last hundred years. In fact, walking it on foot still presented kind of a terrifying prospect.

Zack said, and not for the first time during our outing, that she has one of the best jobs in the world.

"How many other jobs do you get to drive on roads like this? See views like that? Be a part of making the environment better, making the land usable again for wildlife, for people," she said. "And making it safe for people who have got no business sticking their heads where they don't belong."

She explained that holes opening up in the ground due to old mining is just part of the experience of living in Colorado. In a very real sense, Zack's job is to clean up after those who came in the wake of the state's gold rush. It's a job that she's passionate about, but she doesn't blame the original miners for how they left the land. She believes they were just a product of their time.

"Did doctors smoke on TV? Come on—we're talking about

toothless old guys who basically lived here and often inside the mines and didn't know any better," Zack said. "They were just trying to make a living and live past the age of thirty."

And so, as the snow melts and winter recedes to only the base of the highest peaks of the Rockies, the past rears its head again and again. The spring melt off causes sinkholes to open or old tunnels to collapse. Zack comes out to evaluate the issue, finds a solution, and then works with contractors to mitigate yet another mine. As the population of Colorado continues to grow, her office is responding to more calls than ever as landowners are building and coming into contact with old mining areas.

Even if all the mines were discovered and all the mines were closed, she said there will always be work to be done and ongoing maintenance of existing projects. She added corking the hole caused by a mine doesn't necessarily solve the problem.

"So basically we're plugging a tear in the pantyhose with the clear nail polish—but it is still kind of running uphill," Zack said. "Short of filling the entire county with grout underneath—there's no way we're not going to have stuff pop up over and over and over."

And so that's what she does, going from one hotspot to the next, keeping an eye on old projects, finding new ones, and forever fixing whatever issue that crops up next.

CHAPTER 18

BAT COUNTRY

So what can you tell me about the boots?" I asked the room full of clothing-optional enthusiasts. It was late afternoon, I'd been driving for hours, and I had to know. There was a moment where several volunteers at the hot springs looked around at one another, unsure how to respond. One man shrugged and admitted he had no idea, adding that the shoes have always been there as long as anyone could remember.

I had found the mysterious footwear along a lonely stretch of road at the northern end of the San Luis Valley, cutting between acres of golden farmland. The road leads to the base of the Sangre de Cristo Mountain Range for some miles before reaching the hot springs resort. On the southern side, there's perhaps a mile of fencing with posts topped with hundreds of old cowboy boots and shoes.

At first glance, the tattered footwear slowly rotting in the dying light brought to mind unfortunate images of late '70s horror movies and cheery cannibal family traditions. In the surreal late afternoon light of the valley reflecting off my vehicle's windshield, the site did come across as mighty peculiar.

"You would have noticed if any of your guests went missing?" I asked the group. Many looked a tad uncomfortable wearing clothes, which was apparently required in the visitors center, or possibly at my line of questioning.

In point of fact, the unusual display of boots is something of a tradition dating back generations. In the time-honored Western practice of both paying tribute and letting nothing go to waste, worn-out boots were sometimes retired to serve as protection for the tops of fence posts.

In this way, the boots continued to serve their owners by keeping water from destroying the wooden posts. While unusual to see without a complete and helpful understanding of the tradition, it is effective and a form of early recycling. As it turned out, I was at the clothing-optional resort not to partake in the area's well-known hot springs but to pass through it along an old road leading to one of the most efficient and important examples of natural recycling in the state. From the time of the gold rush, there are some thirty thousand mine-related openings across the state today. The bats in Colorado have followed suit by repurposing abandoned mines for their homes.

As the sun set in an ominous red and orange splash across the dreamlike landscape of the valley, distant rainstorms shouldered their way ever closer as I began my two-mile hike. If I hurried, I could see the largest known bat population in the state emerge from an abandoned mine called the Orient.

Very Bat Behavior

I wasn't disappointed when I got to Tina Jackson's office in Denver. A taxidermied bat hung from the wall, rubber Halloween bats hung from the ceiling, and stuffed animal bats were scattered around.

Jackson is a species conservation coordinator with Colorado Parks and Wildlife who is responsible for keeping an eye on the state's bat populations. As it happens, former mines across Colorado are essential to maintaining bats, which have a very direct correlation to the people in the state, their economy, and health.

This relationship began once the gold rush and the fever to burrow into the mountains for precious minerals slowed, and the sounds of picks and rock drills went largely silent. The mines didn't stay that way long; a variety of bat species soon began setting up their hibernation havens and roosting spots.

Jackson said as the state began closing the mines in the early '90s someone realized that bats were living in them.

"Closures make a lot of sense. They are for human safety, which definitely should be done," Jackson explained. "But these mine openings have been around for potentially a hundred-plus years—and the bats have figured that out."

Bats, mountain lions, bears, rodents, and a host of different types of wildlife began making use of what were essentially man-made caves. The state decided to determine which mines were providing essential habitat and which weren't. Bats were one of the animals that the state wanted to encourage in their use of old mines. As such, special gates were installed; like giant Venetian blinds they allowed bats to fly in and out and only restricted humans from entering. Jackson said that some mines are too dangerous even for bats to live in, due to poisonous gases or other hazards, but often they find a way to adapt.

"We have Townsend's big-eared bats that use the uranium mines over in the southwest, so you kind of would think that would be a bad thing, but they're using them," Jackson said. Some believe that the bat populations finding and using mines is a result of their species being pushed by humans into new areas, but Jackson believes the bats appropriating the abandoned mines is more of an opportunistic move than one forced upon them.

"We've just provided a new place," Jackson said. "It was a lot like building a birdhouse. We didn't cut the bird's tree down and then put the birdhouse up; we just put a birdhouse up and now there's another bird that can go, 'Oh, wow, here's a home.'"

"And the bats have certainly jumped on that," Jackson said. A survey of more than six thousand mines in Colorado, Arizona, California, and New Mexico shows that anywhere from 30 percent to 70 percent are being used by bats.

According to Jackson, the Townsend's big-eared bats are the ones most often seen by state biologists making use of the abandoned mines. The big-eared bats are also a species of "special concern" to the state because they are so sensitive to human disturbance. The gradual closure of abandoned mines and loss of caves for bat habitat is thought to have largely contributed to the declining populations of this species in the United States. Jackson added the other poster child of mine-living bat species is the opportunistic Brazilian free-tailed bat, also known as the

Some 250,000 bats fly over the San Luis Valley from the Orient Mine at dusk. The mine has the largest known bat colony in Colorado. *(Photo by Ian Neligh)*

Mexican free-tailed bat, or *Tadarida brasiliensis.*

"That one is using abandoned mines in the San Luis Valley and there we have a bachelor colony of a quarter of a million bats in one abandoned mine," she said.

That same mine is called the Orient, and I was about a quarter of a mile away when I smelled it—long before I reached it. The massive size of the colony can be measured by the smell of guano. Like an old-time restorative of cruel smelling salts, the bouquet of some 250,000 bat droppings is quite powerful. The air becomes heavy with a smell like ammonia, urine, and something akin to moss.

"They're hugely beneficial," Jackson said of bats, and she's not kidding. They are the only flying mammal on the planet to catch and eat some six hundred mosquitoes an hour and put a serious dent in hazardous crop-eating insects. The bats living in the Orient Mine eat two tons of bugs every evening, including the *Heliothis* moth, which makes victim of a host of crops including everything from corn to pumpkins. And the serious dent they put on mosquito populations is not a bad thing either with the rise of charming mosquito-borne illnesses such as dengue, West Nile, and the Zika virus—to name a few.

"Bats are super beneficial. There's been some research recently that says billions of dollars every year are saved," Jackson said of the bat contribution. "They eat agricultural pests, they eat mosquitoes, they eat all sorts of things—and mosquitoes are important from a human disease vector issue."

Protecting the Batcave

The small party of resort goers and myself reached the top of the climb, with a view of the massive cave-in on the upper levels of the Orient Mine. The iron mine was started in the early 1870s and, while not specifically related to gold, is the largest known example of a mine repurposed by bats in the state.

In my time traversing the mountains and talking with prospectors, I've come across more than a few mines specially sealed off with a grate-like structure to allow the passage of bats. But nothing I've seen is on this scale. Too large to block off with a small gate, the "Glory Hole" falls away dangerously into the darkness of the mountain. A giant gaping mouth appeared after the mine closed in the 1930s and silently yawns out over the San Luis Valley. The valley is known for its unusual, often jaw-dropping features such as its colorful gator ranch, UFO Watchtower, and its Great Sand Dunes, but nothing could prepare me for what I was about to see. As the sun sank in what can only be described as an apocalyptic smorgasbord of colors, we waited, whispering only when necessary. They would soon arrive.

Jackson told me protecting mines from humans was essential for the bats' survival. "Waking up a colony of hibernating bats, if it happens often enough, can cause them either to abandon the site or actually die from not having enough fat reserves to make it through the winter," Jackson explained. "So we're protecting the people from the mine, but we're also protecting the bats from the people."

Biologists will go to a mine site eligible for closure and look for signs of bats, such as guano or bug parts, littering the ground. Failing that they'll wait for winter and do their best to sneak inside, which is always a dangerous proposition in an abandoned mine.

In 2010 the state worked on nine abandoned mine projects involving bats. These projects consisted of 118 mine openings. That year fifty-five bat gates were installed. In Clear Creek County some thirty-six mines near Dumont were investigated for bats. Of those, twenty-two mines were closed and fourteen received bat gates.

Jackson said white-nose syndrome, a malignant fungal disease that eats away at hibernating bats and brings with it a high mortality rate, hasn't yet reached Colorado. Biologists are regularly and vigorously monitoring for it in Colorado's abandoned mines. The fungus was first observed in 2006 and has since affected at least nine species of hibernating bats and spread to twenty-nine states.

And there are a lot of abandoned mines in Colorado that still need to be checked and closed every year. Of the roughly thirty thousand mine-related openings across Colorado, ten thousand of them have been made bat friendly in the last twenty years.

"So we're looking at about another forty years before we get to the remainder of them," Jackson said. "But on the other hand we're building closures, but those fail at times. Those fail because people make them fail."

Indeed, the locations of mines with bats living in them is something of a secret. I'd seen a few, but they were pretty far off the beaten path. Jackson said work to repair and maintain bat gates is a never-ending process. She added that a mine with bats on private property is never fully off-limits.

"When gold prices get high enough . . . it makes sense for people to open them back up," Jackson said. "We have had cases where our biologists go up to survey, and the gate is open, and there are people working on it."

Staring up at the sky I spotted two bats fly overhead. Then a few more, then hundreds came twisting out of the mouth of the mine like a single living thing. The bats flew in a thick black column, not far above where I stood, stretching in the cool evening air like a snake. They made some sounds, the flap of a wing or a random squeak, but mostly remained silent in their single-minded purpose to drift over the valley. Thousands of bat wings sounded not unlike the wind whispering over the ocean.

For nearly half an hour bats got in line and took their turn entering the giant formation, leaving the mine for their nightly hunting grounds. It seemed to me that it was a sight as valuable as any gold or other precious metal ever removed from the mountains of Colorado.

CHAPTER 19

GHOST COUNTRY

Zigzagging, crooked, slanted, broken. Maybe a dozen buildings remain. The town of Independence lies largely forgotten in the shadows of mountains surrounding it like steepled fingers. Wooden splinters from abandoned cabins reach out to the cloud-studded sky like the broken masts of ships. Nearby, Roaring Fork River babbles to no one in particular, and Highway 82 sneaks its way far above and out of sight through a troubled mountain pass to the town of Aspen.

Traffic, cautious of the winding grades, cliffs, and sudden merging lanes, moved along without any awareness of the former mining settlement below. Soon the pass, situated at 12,095 feet as it tap dances along the Continental Divide, would close for the winter.

Nearly ninety buildings once existed in this town. Depressions in the soft earth show evidence of old dwellings. The remains of many of the town's still standing structures lie open to the sky, like the ribcages of long-dead animals. The wood now looks like bones, bleached by the wind and relentless cold hammering down from the mountain peaks. This would have been a difficult town to live in during the best of times, but gold drew people to the sides of mountains to cling to civilization in search of fortune. Today this ghost town is truly more skeletal remains than spectral in nature.

The town of Independence was formed after gold was discovered in 1879. It once had nearly fifty businesses, including seven restaurants, two boarding houses, three post offices, a newspaper, and several saloons. *(Photo by Ian Neligh)*

It was still technically summer but in the high country, the bushes were beginning to turn yellow and orange with the promise of an early and long winter. The buildings lay scattered like brown dice across the valley, dusted in places with a stubborn red plant, covering the ground and mountainsides like rust. One would have to be stubborn and tough as nails to live in a place like this. The story goes the site was first discovered on the Fourth of July—hence the name Independence—and once sheltered more than a thousand people.

Independence was a gold town deep in silver territory. Burton Avenue, Sixth Street, Johnson Avenue—the town's original map shows the early grid-like beginnings of a fully realized town, and twenty buildings once sat shoulder to shoulder along Aspen Avenue. It's said that the best way to identify the locations of buildings no longer standing is to look for holes in the ground, signifying the remains of privies.

This was once a place wealthy with gold but eventually the payouts became too few and the cold too much. This town struggled for breath in the thin air and died with only the pine trees bearing witness. Independence was discovered in 1879—twenty years after the first gold rush. It is safe to say that Colorado had several gold discoveries that blasted headlines across local newspapers and sent people up ever farther into the Rockies for gold.

First, a desperate tent city, then a few small cabins signified the town's start. The gold in the area was rich and $100,000 was found in the first year alone. For many, this kind of money justified the suffering at the hands of Mother Nature, but four walls and a roof were essential to any sustained living in this type of environment. Soon as many as fifty businesses set up shop to cater to the miners and their families. The area's gold mines were combined by the Farwell Consolidated Mining Company in 1881. It is believed that Independence once had seven restaurants, two boardinghouses, three post offices, a newspaper, and a host of saloons.

But like all gold towns the mineral was destined to soon run dry, like the promise of water in a high mountain desert. The veins petered out and the wealth that once brought in investors, businesses, and families was all too quickly gone. The residents dwindled to one hundred, but the town struggled on, always hopeful that more gold would be discovered before long.

The coup de grâce came in 1899 when a series of violent snowstorms battered the state and utterly cut off supplies to the town. With food scarce and the threat of avalanches high, the town's residents decided they had to leave to survive. And the only way to do that was to ski out.

Special correspondent to the *Aspen Weekly Times* and resident of Independence, Mrs. D. H. Elder reported to the paper that by February 2 many of the remaining residents were prepared to leave.

> This camp, which one week ago had nearly one hundred men working, and the large 30-stamp mill and concentrator, handling 70 tons of ore per day, is today depopulated. . . . It is safe to say that seventy-five pair of Norwegian snowshoes or "(skis)" . . . have been made in the last three or four days and by tomorrow evening our camp will be almost deserted except by families who are forced to stay for transportation out. Even old snowshoers like the writer are loth [sic] to venture out while the

> snowy caps are toppling upon the crest at either side and threatening to engulf the valley at any point and at any moment! It is truly a reign of terror; and not until one has been situated where they momentarily expect to hear the telltale rumble of the monster which would crush every home in this valley as quickly and as completely as one could crush an egg shell in the hand, can one realize the full and awful meaning of the words "impending danger." To add to the general terror of the occasion the wind roars and howls about the houses, completely enveloping them in such blinding clouds of whirling snow that for five minutes at a time one cannot tell whether the other little homes surrounding us remain or whether they have succumbed to the fury of the elements.

Elder said some of the area's men broke into a package addressed to the woman who operated a local store and took what they wanted, leaving the rest to lie in the snow and at the mercy of the weather. The town, trying to keep a sense of humor given the bleak and dangerous situation, decided to call the mass exodus from Independence to Aspen a race—the entry fee being a ham sandwich, which was to be eaten along the way. The next day Elder noted that the site of so many people trying to get around on and unfamiliar to wearing skis was quite humorous.

> Your correspondent has got so she can take a long breath without expecting to jar a snow slide loose, and possibly we may retain sufficient nerve or foolhardiness to remain with the camp. If it were not for the extreme seriousness of the occasion, there would be many very ludicrous situations to write up, as a result of the first efforts of some fifty to seventy-five inexperienced men, making and learning to master the Norwegian skies. All are not experts in either line and every variety of wooden skates and snowshoe footwear is being made and tested preparatory to the flight out of the gulch.

She wrote when the time came to leave on February 3, nearly one hundred people were on the road, some helping others to move livestock and horses through the deep and trying snowbanks. "Others [were] seeking their own safety in a desperate rush to get out of the threatening dangers which seems to so completely surround us. To remain seems to be almost foolhardy and to attempt to get out is still far more so, as the danger of slides along the road is even worse than that which threatens here."

Elder stayed as the others left to travel the sixteen miles to Aspen. A group of men leading some twenty horses stopped to catch their breath and eat a small dinner around a fire. The avalanche struck around midnight.

> The boys were huddled close to a large rock eating . . . when the telltale rumble warned them of the approach of the slide. [One man] says he had just enough time to say "We are done for boys!" when the slide struck them under about four feet of snow. Some of the boys were carried completely over the rocks, but as they were in the edge of the snow they managed to dig themselves out unhurt, while the horses less than twenty feet farther in were so completely crushed with the weight that some . . . were dead the moment it passed.

In fact, all but two horses were killed by the avalanche. One horse was swept across the length of the valley and left uninjured on the opposite side of the gulch.

While there was talk in the newspaper from the mine's officials about returning and starting up operations again, it just wasn't meant to be. Some, however, did stay behind, and conditions continued to worsen. The paper's special correspondent, Mrs. D. H. Elder, appeared in its pages again about a month later on March 25, when she and another woman skied down into Aspen to collect supplies and return to Independence.

> Mrs. T. L. Hopkins and Mrs. D. H. Elder, the two brave Hunter Pass women, arrived from that snowbound camp yesterday, having made the trip out, eighteen miles on Norwegian snowshoes (skis). Mrs. Hopkins, when about half way down, met with a painful accident which delayed them somewhat. In coasting down a hill, the lady got a hard fall which severely sprained a wrist. Besides being painful, the injury prevented her from using the pole which is an essential part of the snowshoer's outfit. But they finally reached town all right, and after a day or two's rest will make the return trip. Mrs. Elder says the snow about Independence is becoming

The residents of Independence were forced to evacuate in 1899 from fear of avalanches. *(Photo by Ian Neligh)*

> granulated, and small slides are now quite numerous. The trip, she says, is quite dangerous now, especially during the part of the day when it is thawing.

Try as the miners might to keep the town alive, the final death rattle was already deep in its throat. Colorado has a countless number of ghost towns born in the dazzling flood of those looking to strike it rich. Most didn't experience the trauma of having its residents forced to flee from the danger of avalanche and starvation. Most, in fact, died the ignoble death that comes at the hands of economic strangulation. The mine dries up, and its miners leave for other locations. The merchants follow and soon all that is left behind are sagging wooden walls and empty streets. Many dangers were, however, associated with living in the Rockies and some towns were in fact wiped off the face of the map—or, in the case of Brownville, buried alive.

Brownville

While gold was initially sought, the prospectors who found themselves on the west end of Clear Creek County discovered silver. They were uncertain of what to do with it. But once the riddle of silver was solved, fortunes and towns were created almost overnight. In 1860 a miner discovered wealth west of Georgetown and in ten years' time Brownville was founded. The town was said to have grown to the point where it had eating establishments, saloons, and other mining-town essentials.

Like every mining town, its residents would head out at the crack of dawn to work in the area's many mines, and then return at dusk bone-weary from digging, shoveling, and blasting through the hard rock like termites through wood, ready to start all over again the next day. Located above the town of Brownville was the famous Seven-Thirty Mine, so named for owner Heneage Griffin's reluctance to blow the whistle at the usual time of 6:30 A.M. Griffin was considered by one mining journal in 1891 to be one of the wealthiest men in Colorado.

Born in England, he was educated at Christ Church at the University of Oxford before deciding to come out west and seek his fortune, which he found in Colorado. In time, he decided to have his younger brother Clifford sent for and brought out to work with him. But Clifford had a dark past, which haunted him. The story goes he was to be married to a beautiful woman who was found dead in her room the night before their wedding.

Clifford traveled to Colorado and worked at the mine as its superintendent. He was liked by the miners and was known to play his violin beautifully up at his cabin, near the mine, above Brownville. His music would drift to the town below, attracting many residents to come out to listen, occasionally giving applause. But the music he played was said to be mostly sorrowful, not unlike the man himself.

The tale goes that after one summer night's performance, the music ended with the sound of a gunshot. People reportedly rushed up to find the young man dead from a bullet to the heart, already lying in a grave he dug from solid stone, with instructions to leave him there. This colorful bit of local lore loses a bit of luster when looking at the coroner's inquest written in June of 1870. It stated that Clifford was discovered in his cabin, not in a grave hewn from granite, but with a self-inflicted bullet wound to his head. He was, however, buried in a spot just outside his cabin, and a spectacular monument was erected by his older brother. Those with a morbid inkling can hike to the memorial to this day.

Meanwhile, the mine's dump had accumulated and grown larger for years. During one spring flood, thousands of tons were sent tumbling down from the mine to the town below. This would happen again and again until it was no longer safe to reside in Brownville. In 1892 one such slide came down and covered several homes and the saloon. A few years later in 1895, another slide destroyed several more buildings. That slide was followed by another later that year said to have taken out Griffin's cabin and mining buildings, filling the town with debris. Whether from poor management of the mine's dump or the ghost of Clifford Griffin, the town's citizens must have had years of uneasy evenings thinking about the rocks that could come tumbling down onto them at any moment.

In 1912 the residents heard again those familiar rumblings from the surrounding hills and evacuated the town and nearby Silver Plume. Once more, the rock and dirt came down and destroyed, damaged, or buried houses, businesses, and the town's hotel. Following the massive debris came mud, which is said to have slowly entombed the town as its former inhabitants watched from safety. It was probably at this point they decided to toss in the towel and move elsewhere. Today much of Brownville lies buried beneath Interstate 70. Most traces of the town are gone.

Brownville was a victim of its mining success. Other towns in Colorado were also so precariously situated that its residents only escaped instant death by sheer luck and being in the right place at the right time.

Masontown

In 1866 a gold mine was discovered about twenty-three miles southwest of Brownville as the crow flies. The man who found the valuable resource built a road to access the area, and soon houses grew like weeds around it. The mining settlement was located in an area even more precarious than the site of Independence, clinging only by its fingernails to the side of a mountain.

In 1912 or the early '20s, the date varies, the denizens of Masontown decided to travel down to nearby Frisco to enjoy an annual winter party. They were basking in revelries when the thunderous echo of a massive avalanche interrupted their celebration. With the devastation of Thor's hammer, the slide swept most of the hamlet clean off the side of the mountain, destroying most of its houses and buildings. The avalanche occurring when Masontown's residents were gone was no doubt something of an early Christmas gift, but it also signaled the end of this mining settlement.

The Winds of Change

It's impossible to know for certain how many towns flickered and died in embers during the gold rush. A few cabins and a saloon could grow into a town or city of more than a thousand and still wither and die before, or just after, the turn of the century. There are dozens of books on the subject of Colorado ghost towns, but even they only scratch the surface. There's no way to appreciate the number of towns once fighting to get a foothold on Colorado soil only to be blown away by the winds of economic change or instability.

Some can still be found in the remains of fireplaces or dried-out log homes. Others still have maybe a handful of occupants glaring out Victorian windows at curious onlookers. The gold town of Tin Cup in Gunnison County, once considered one of the most dangerous towns in Colorado, is just such a place. The town had four cemeteries, a staggering twenty saloons, some three thousand residents, and has suffered a perpetual hangover these past one hundred years.

Its first marshal quit when he wasn't paid; its second was fired when local officials thought he'd be lynched. The third was shot by the second, and the fourth was shot by a gambler. The fifth quit to serve God, the sixth was committed to an asylum, and the seventh was also shot. The eighth marshal of Tin Cup finally managed to last the full length of his assignment. The town slowly withered away and exists today largely due to a steady stream of tourists.

With houses like wooden mausoleums, ghost towns pepper Colorado's Rockies. There's Animas Forks, Ashcroft, Capitol City, Caribou—more than six hundred remain in some fashion. Most towns labeled as a Colorado ghost town share a fate more similar to that of Tin Cup than Brownville, Independence, or Masontown. But in time it wasn't only the towns that suffered the economic whims of the gold mining industry, but the industry itself.

Gold came and went, and so too did those who came to look for it since the precious metal's discovery led to the westward rush in 1859. The state's gold mining industry gave way to silver production and took the crown back again during the Great Depression. It was World War II that nearly extinguished gold mining in Colorado and gave it a wound from which it would never recover.

Nonessential

In communities like Gilpin County, mining had essentially slowed and died by 1910. Additionally, the miners had dug so deep into the mountains that it became too costly for work to continue. For example, in the fall of 1918 an ounce of gold was worth $20.67 but cost as much as $30 to produce. As mines closed, the populations of many mining towns drifted away, leaving schools and businesses empty. With the Depression of the 1930s came a flood of cheap mining labor, and for a time mining for gold made a comeback.

The bombs that fell on Pearl Harbor on December 7, 1941, reverberated across the country and the subsequent call to fight reached many of the Colorado mining towns. Men enlisted in terrific numbers and mining production decreased to where it simply wasn't feasible to work many of the mines, and those that still functioned could not obtain essential materials. Gold mining took one hit after another like a used-up boxer before the final blow was delivered, knocking the industry to the mat for good. In 1942 the War Production Board issued Order L-208, closing gold mines across the nation. All minerals considered nonessential to the war effort had to cease production. The majority of those mines closed and never opened again. Some mines in Gilpin County are reported to have mining equipment and vehicles from the 1940s stored and secured for the day when the war ended and its owners could return and resume work. They never did, and the mines stayed shut.

Some gold mines did begin production again after the war but by

1967 Colorado dug only twenty-two thousand ounces of gold from its mines, the lowest production year since George Jackson's discovery caused that desperate rush west. Towns that once relied solely on its miners and minerals had to look to other sources of revenue. Many mining towns began looking to tourism and skiing traffic to continue their tenuous existence. While the gold may still be locked away in the mountains, the time of large-scale mining in the Rockies is now predominantly relegated to the pages of history. But as ever, the gold still calls to some and there are those who still answer.

Epilogue

My eyes strained for hours to find it. The gold was small and, in retrospect, likely not worth the sweat, pain, and misery poured into discovering it—and yet I wanted more and was delighted with what I had. That day I'd already spent close to five hours huddled over the river, the steep canyon walls providing little shade from the sun, which made droplets of sweat fall into the river beneath me. Despite the hour I wanted to go for another five hours, five days, five years.

For nearly twelve months I'd stepped, however briefly, into the shoes of those who had shaped the West as we know it today, and the experience was equal parts exhilarating, astonishing, and weird. After all, I'd found gold.

I sensed, or maybe heard, someone behind me. Not quite delirious from squatting over a river for the better part of a day, but certainly flirting with dehydration, I turned around to see a grinning Japanese tourist. The sight was several shades of surreal. Then the man said something that I couldn't quite make out. He repeated himself, "Gold—finder?"

It took me a few moments to emerge from my gold-induced stupor to answer his question, but finally I nodded and said, "Yes." I had brought along a visiting friend to demonstrate what I'd been up to these long months, and together we showed the tourist the vials of dark soil that we'd collected. In the vial was gold, which we'd scooped up with the black dirt. The man came close, eager and interested at first in our discovery. He cautiously peered into the glass tube of collected soil. Unconsciously a frown began to form on his face as he continued squinting into the dubious mixture harvested in the little bottle.

I had a pretty good idea what he was thinking, and the answer was "yes." As thrilled as I was to find gold, it was indeed a woefully small amount. The miners and prospectors that I'd interviewed wouldn't have wasted their time and energy on such a small quantity. Or if they had, it would have been to add to many other similarly sized samples.

Deciding perhaps he'd come upon some wayward buffoon, I could see he made a quick decision, nodded politely, and walked to the edge of the stream where I had spent the majority of the day. He then held out his hands in a gesture that seemed to say, "'Look at this, this is beautiful—and maybe this is something we can agree on?"

I knew I needed to up my gold-hunting game. I needed to go back to the people I'd interviewed to ask them how they'd done it.

Six Hundred Feet Underground

I followed the man who would rather be anyplace else in the world than leading a small group of journalists into the former gold mine for the historic meeting. And that was OK with me, so long as he didn't go there before showing me the meeting location—and then was kind enough to show me the way out again. It was the winter of 2015 and I tried, again and again, to keep from scuffing the top of my helmet on the low-hanging ceiling of the Idaho Springs Edgar Mine. In this, I was unsuccessful and sighed in undisguised relief as we got to our destination.

Haphazard cords for camera lights were strung across the dirt floor in an open chamber and cold water dripped from the ceiling. It was the first time a United States Congressional hearing subcommittee met more than six hundred feet below the surface of the planet. US Representatives Cresent Hardy, R-Nevada; Doug Lamborn, R-Colorado; Rob Bishop, R-Utah; Ed Perlmutter, D-Colorado; and a host of mining experts and officials sat in folding chairs with literally a mountain of rock suspended above their heads. These were the members of the Energy and Mineral

Resources Subcommittee, and they had found a solution to the county's rapidly declining number of mining engineers.

Before beginning the hearing, one senator leaned forward in his chair, across the table, and candidly said: "First of all, I just gotta say, this is weird." He was right; this was a strange place to have a meeting—but given the topic was the future of mining, it was also perfect.

His comment elicited an awkward round of chuckling. In the room were maybe thirty people, and reporters were segregated off to one side in a special journalist zone, which was ludicrous as there were maybe three of us. An uncomfortable-looking young woman and a couple of guys manned an unwieldy camera, baring their teeth at anyone who came close enough to endanger their equipment. I got up, ignoring the meeting's decorum, and walked closer to see the several middle-aged men and women preparing to discuss the future of mining.

Their goal was to talk with mining industry experts, receive their testimony, and ultimately introduce, and try to pass, the Mining School Enhancement Act. The act would pump much-needed funding into efforts to improve mining education.

This need for meeting underground and talking about funding for better education had come about after the Gold King Mine spill, which had occurred only some months before. The Environmental Protection Agency (EPA) caused an accident in the old gold mine that unleashed some three million gallons of contaminated, bright orange water into the Animas River. The river turned the unsettling color of a peach milkshake, and what followed was a flurry of environmental damage and lawsuits against the state and the EPA by both the state of New Mexico and the Navajo Nation.

The Energy and Mineral Resources Subcommittee did an investigation and found that the EPA had a severe deficit in the number of mining engineers and qualified geologists working in the affected area. The group agreed that going forward more experts trained in mining across the country were needed to avoid future environmental catastrophes.

The meeting ended and I was ready to get out of the mine—they are comfortable only if you're hobbit sized. As we solemnly marched out and back to our vehicles, I thought about how the mining industry had dwindled as well as the number of people educated to manage the issues associated with it. The subcommittee estimated that in the next decade about 70 percent of the leaders of the country's mining industry are expected to retire. Even if we never see a return to mining's glory days in the West, it is essential to have those equipped with the proper knowledge ready to deal with the issues we've inherited.

The Handful

As I tracked down the small-time miners still carving a living from the Rockies, I found that there weren't many of them left either. Both large- and small-scale gold mining aren't even shadows of what they once were. Recreational gold prospectors and miners will always see their ranks swell and dwindle depending on the price of gold, but those who do it full time, and legally, amount to only a handful. When asked if there was someone I had missed in my research of current day gold miners, Al Mosch best summed up the answer by grinning and asking, "Do you have a Ouija board?"

One day I ran into Ken Reid sitting outside a pizza restaurant in a mountain town, and after talking for a bit he admitted he was scheduled for back surgery. Years of being hunched over rivers hadn't done his spine any favors. After a lifetime of looking for gold, he could no longer pan or climb into a wetsuit and had recently sold his claim. His dream now was to move to Australia where he would use a metal detector to look for gold. He also admitted, somewhat sheepishly, that he'd bought a nice guitar, which he could never previously afford because all his money went into his mining operation.

After a year of research and interviews, things were changing and so too was the weather. I'd started in the winter sitting in Reid's "Man Cave" on the banks of Clear Creek over a bucket of ice-cold water as we panned through his concentrates gathered the summer before. Now the aspens were turning yellow again, fluttering in the cooling breeze. The streams were getting low and soon would be slowing with the promise of ice. Fall had come early and promised a severe winter, and the streams and any precious metal they held would be locked away for another season.

I wanted to try to find gold one more time. Not a microscopic fleck,

Gold miners Chad Watkins and Jesse Peterson evaluate the potential for gold in the Chaffee Mine. *(Photo by Ian Neligh)*

but something significant. I wanted to find a piece of gold you could see with the naked eye without giving yourself an aneurysm for your trouble. I asked to meet once again with two of the most accomplished gold panners still working in the Rockies. On a Friday morning I met with Chad Watkins on Jesse Peterson's property for one more chance to learn about looking for gold, maybe learn something that I had missed.

I pulled onto Peterson's property in Gilpin County, got out of my truck, and greeted one of the laconic old-timers that works with him. As he tried to track down his boss with a series of text messages, he lit a cigarette and told me that it's been a slow year for tourists looking to do a little gold panning. "Not like last year," he drawled.

I met Watkins and Peterson by the mouth of the Chaffee Mine. We walked around inside a bit to see what progress Peterson had made in preparing it for operation and what areas he wanted to explore. Watkins told me he was looking to get into hard rock mining himself and maybe leaving the river to buy his own gold mine. He took particular interest in the readings of Peterson's mine with a small hand-sized metal detector as we went ever-deeper inside.

For my part, I decided I wasn't a huge fan of mines. I spent a good amount of time in them, and with every story I wrote or came across dealing with some hideous mining accident, my enthusiasm to spend more time inside significantly waned. We continued walking and soon the wooden ties, where iron track once lay, became waterlogged, half sunk in mud and rough-hewn. Peterson told me they were the original 1870 mining cart ties. He was literally picking up where the original miners had left off more than 140 years ago.

On the return trip, I was the first to step outside, while the other two lingered. I was out there maybe a minute when I happened to notice a bowling ball–sized rock come bouncing down from the mountainside above the mine. About twelve feet away from me it caught air and whistled past to the ground with a resounding thump. It then occurred to me that it might have been safer back inside.

Gold!

Soon I found myself working with the two, learning the techniques of gold panning in a more nuanced way than I had previously employed. Peterson said most people go out and try panning and fail, return home, and put the pan on their kitchen, eventually filling it with fruit.

He was working on a much larger scale than I, but said I couldn't proceed to more advanced gold mining techniques until I had mastered the basics. As an example of how to do it, Peterson panned down a portion of dirt, discovered a gold flake, dropped a massive fistful of dirt and muck back on top of it, and told me to find it again. Learning gold panning from him was like learning kung fu from a Shaolin grandmaster. I worked at it until I found the hidden flake of gold. Then I found another and another. I felt the strong tickling of gold fever.

"That's it, good job," Peterson said, eyeing the most recent gold I'd panned out. I thanked them both for the lesson and headed back to my office. I was still thinking about gold when my eyes rested on a fist-sized rock sitting on my desk that Al Mosch had given me when we'd first met. He had been driving in front of me, leading the way to his mine, when he stopped his car suddenly and reached down to grab the rock up off the road. He proudly handed it to me later and said the piece was an excellent example of local gold ore. I'd hung on to it ever since.

I thought, in theory, I could take the rock, put it in a towel, and smash it with a hammer until there was nothing left but a fine dust. I could then put the remnants into my gold pan and see if it indeed contained any gold. And I was tempted. How much gold would be in the rock? What if I could find as much as an ounce? If I didn't pulverize it, I might never know. I considered that there might not be any gold in the rock, that it might just simply be a rock. Then, after a long moment, I decided to leave it alone.

—Ian Paul Neligh
Idaho Springs, Colorado

Special thanks to the Historical Society of Idaho Springs and the Gilpin Historical Society. A huge thanks to all those who let me follow them around for the better part of a year.

Also, thanks to Dave, Deb, Dan, and Eva. You can't write a book like this without relying on the authors and journalists who have come before. It is on their shoulders that we all stand.

Bibliography

Adams, Phyllis. "Gold Mining Wins and Loses in the 20th Century." *Clear Creek Courant*, September 28, 2009.

Athearn, Frederic J. *A Forgotten Kingdom: The Spanish Frontier in Colorado and New Mexico, 1540-1821.* Denver, CO: Bureau of Land Management, Colorado State Office, 1992.

Bancroft, Caroline. *Gulch of Gold: A History of Central City, Colorado.* Denver, CO: Sage Books, 1958.

———. "The Elusive Figure of John H. Gregory, Discoverer of the First Gold Lode in Colorado." *The Colorado Magazine*, July 1943.

Brown, Robert L. *Ghost Towns of the Colorado Rockies.* Caldwell, ID: Caxton Printers, 1968.

———. *Colorado Ghost Towns Past and Present.* Caldwell, ID: Caxton Press, 1972.

Colorado. Bureau of Mines. *Thirteenth Biennial Report of the Bureau of Mines of the State of Colorado.* "The Argo Tunnel" By Rens Schirmer. Denver: Smith-Brooks Printing, 1914. https://archive.org/stream/biennialreportis00colo_0/biennialreportis00colo_0_djvu.txt.

Dallas, Sandra. *Colorado Ghost Towns and Mining Camps.* Norman: University of Oklahoma Press, 1985.

Davis, Mark W., and Randall K. Streufert. *Gold Occurrences of Colorado.* Denver, CO: Colorado Geological Survey, Dept. of Natural Resources, 1990.

Dugan, Ben M. *Mines of Clear Creek County.* Charleston, SC: Arcadia Publishing, 2013.

"Evidence of an Enormous Mineralization." *The Mining American* LXXVII (April 29, 1918): 6. Google Books.

Fenwick, Robert W. "Alfred Packer: The True Story of the Man-Eater." *Empire Magazine* of the *Denver Post,* 1963.

"Fourth Body Taken from Argo Tunnel at Idaho Springs." *Steamboat Pilot* (Steamboat Springs), January 28, 1943, 27th ed.

Greeley, Horace. *An Overland Journey from New York to San Francisco in the Summer of 1859.* New York: C.M. Saxton, Barker &, 1860. https://books.google.com/books?id=v3EFAAAAQAAJ&printsec=frontcover&source=gbs_ge_summary_r&cad=0#v=onepage&q&f=false.

———. "An Overland Journey, New York to San Francisco, Summer of 1859." Edited by W. Ali and Charles Duncan. *The Geographical Journal,* 1965.

Hanchett, Lafayette. *The Old Sheriff and Other True Tales.* New York: Margent Press, 1937.

Historical Society of Idaho Springs. *History of Clear Creek County: Tailings, Tracks, & Tommyknockers.* Denver, CO: Specialty Press Inc., 1986.

Hollister, Ovando James. *The Mines of Colorado.* New York: Promontory Press, 1974.

Ivy, Logan D., and James W. Hagadorn. "Exploration of Colorado's Deepest Roots." *Denver Museum of Nature & Science Annals 4* (December 31, 2013): 179-230.

Jackson, George A. "Jackson's Diary of '59." Place of Publication Not Identified: Placer Inn, 1929.

Jarvis, Heather. "Discovering Tom's Baby, the Largest Piece of Gold Ever Found in Colorado." *Summit Daily* (Frisco), December 7, 2015.

Jessen, Kenneth. *Colorado Gunsmoke: True Stories of Outlaws and Lawmen on the Colorado Frontier.* Loveland, CO: J.V. Publications, 1986.

Kinder, Gary. *Ship of Gold in the Deep Blue Sea.* New York: Atlantic Monthly Press, 1998.

Lakes, Arthur. "A Typical Gold Mine." *The Colliery Engineer and Metal Miner* XIV (June 1894): 282-85.

Leyendecker, Liston, Sandra Dallas, and Stephen Mehls. "Colorado Ghost Towns and Mining Camps." *The Western Historical Quarterly*, 1986.

Marotti, Ally. "Gold Leaves for Colorado Capitol Building Dome Return from Italy." *Denver Post,* May 19, 2013. http://www.denverpost.com/ci_23489187/gold-leaves-colorado-capitol-building-dome-return-from.

Mosch, Alvin. *Gold Dust, Dreams & Life of a Miner: True Short Stories & Gold Mining Adventures from the Secret Diary of Alvin Mosch.* Idaho Springs, CO: A. Mosch, 2008.

Murphy, Jan. *Outlaw Tales of Colorado.* Augusta, GA: Morris Book Publishing, LLC, 2012.

Orange, Will J. "A Big Enterprise." *Silver Cliff Rustler* (Silver Cliff), December 20, 1893.

Report by Navo, Kirk W., Lance Carpenter, Amy Englert, Tina Jackson, Chris Kloster, Dan Neubaum, Ed Schmal, Mike Sherman, and Raquel Wertsbaugh. "Bats/Inactive Mines Program: 2010 Mine Evaluations." This project was funded by Great Outdoors Colorado, the National Fish and Wildlife Foundation, the Colorado Division of Wildlife, US Forest Service, and the Bureau of Land Management.

Neligh, Ian. "A Golden Legacy." *Clear Creek Courant* (Idaho Springs), February 2, 2009.

Phillip, Abby. "How Treasure Hunter Tommy Thompson, 'One of the Smartest Fugitives Ever,' Was Caught." *Washington Post*, January 30, 2015. https://www.washingtonpost.com/news/morning-mix/wp/2015/01/30/how-treasure-hunter-tommy-thompson-one-of-the-smartest-fugitives-ever-was-caught/.

Smith, Duane A. *The Trail of Gold and Silver: Mining in Colorado, 1859-2009*. Boulder: University Press of Colorado, 2009.

Smith, Tierra. "Newmont Completes Purchase of Cripple Creek & Victor Gold Mine." *Denver Post,* August 3, 2015. http://www.denverpost.com/business/ci_28579489/newmont-completes-purchase-cripple-creek-amp-victor-gold.

Sowell, Merle L. *Historical Highlights of Idaho Springs Mining Camp Days.* Idaho Springs, CO: Idaho Springs Friends of the Library, 1976.

Stewart, K. C., and R. C. Severson. *Guidebook on the Geology, History, and Surface-water Contamination and Remediation in the Area from Denver to Idaho Springs, Colorado.* Washington: U.S. G.P.O., 1994.

Stiff, Cary. "Clifford Morrison Killed Last Week When Shaft Caves in on Him at Stanley Mine." *Clear Creek Courant* (Idaho Springs), May 27, 1977.

———. "Death at the Stanley." *Clear Creek Courant* (Idaho Springs), May 27, 1977.

Storms, Aaron. "Turning Back the Pages." *The Weekly Register-Call.* Storman Media, LLC, April 21, 2016.

Taylor, Bayard. *Colorado: A Summer Trip.* New York: G.P. Putnam, and Son, 1867.

Tuttle, Merlin D., Daniel A. R. Taylor. "Bats and Mines." Bat Conservation International Inc., Resource Publication No. 3.

Ubbelohde, Carl, Maxine Benson, and Duane A. Smith. *A Colorado History.* 8th ed. Boulder, CO: Pruett Publishing, 2001.

US Department of the Interior Bureau of Land Management. *Abandoned Mines Are Potential Killers.* US BLM. http://www.blm.gov/style/medialib/blm/wo/MINERALS__REALTY__AND_RESOURCE_PROTECTION_/aml.Par.37424.File.dat/BLM_ALM_SafetyBrochure.pdf.

Villlard, Henry. *Memoirs of Henry Villard, Journalist and Financier*. Boston: Houghton, Mifflin, 1904.

Voynick, Stephen M. *Colorado Rockhounding: A Guide to Minerals, Gemstones, and Fossils.* Missoula, MT: Mountain Press Publishing Co., 1994.

Wilder, Frank C., ed. "Great Things for Clear Creek County." *The Mining Investor* 74, no. 5 (1914): 74.

Willoughby, Tim. "High-grading: Stealthily Stealing Silver." *Aspen Times,*

December 10, 2010. http://www.aspentimes.com/article/20101212/ASPENWEEKLY/101219990.

Zamonski, Stanley W., and Teddy Keller. *The '59er's Roaring Denver in the Gold Rush Days: The First Three Years.* Frederick, CO: Platte 'N Press, 1983.

Anonymous. "Colorado Miners Escape Death When Tunnel Is Flooded." *Gazette Times* (Pittsburgh), February 1, 1912.

———. "From Hunter Pass." *Aspen Weekly Times,* February 11,1899.

———. "Georgetown Has a Haunted House." *Rocky Mountain News,* December 26, 1869.

———. "An Immense Nugget of Gold." *Daily Journal* (Breckenridge), July 25, 1887.

———. "The Late Duel." *Western Mountaineer*. Golden, Jefferson County, CO, Wednesday, March 14, 1860, 2.

———. "Our Eureka Ghost." *Weekly Register Call,* January 19, 1874.

———. "Retribution About Reaching the Malefactor." *Rocky Mountain News Weekly* (Denver), August 29, 1860, 19th ed. https://www.coloradohistoricnewspapers.org/cgi-bin/colorado?a=d&d=RMW18600829.2.53#.

———. "Women on Snowshoes." *Aspen Weekly Times,* March 25, 1899.